Network Embedding

Theories, Methods, and Applications

Synthesis Lectures on Artificial Intelligence and Machine Learning

Editors

Ronald Brachman, *Jacobs Technion–Cornell Institute at Cornell Tech*
Francesca Rossi, *IBM Research AI*
Peter Stone, *University of Texas at Austin*

Network Embedding: Theories, Methods, and Applications
Cheng Yang, Zhiyuan Liu, Cunchao Tu, Chuan Shi, and Maosong Sun
2021

Introduction to Symbolic Plan and Goal Recognition
Reuth Mirsky, Sarah Keren, and Christopher Geib
2021

Graph Representation Learning
William L. Hamilton
2020

Introduction to Graph Neural Networks
Zhiyuan Liu and Jie Zhou
2020

Introduction to Logic Programming
Michael Genesereth and Vinay Chaudhri
2020

Federated Learning
Qiang Yang, Yang Liu, Yong Cheng, Yan Kang, and Tianjian Chen
2019

An Introduction to the Planning Domain Definition Language
Patrik Haslum, Nir Lipovetzky, Daniele Magazzeni, and Christina Muise
2019

Network Embedding: Theories, Methods, and Applications
Cheng Yang, Zhiyuan Liu, Cunchao Tu, Chuan Shi, and Maosong Sun

ISBN: 978-3-031-00462-9   paperback
ISBN: 978-3-031-01590-8   ebook
ISBN: 978-3-031-00035-5   hardcover

DOI 10.1007/978-3-031-01590-8

A Publication in the Springer series
SYNTHESIS LECTURES ON ARTIFICIAL INTELLIGENCE AND MACHINE LEARNING

Lecture #48
Series Editors: Ronald Brachman, *Jacobs Technion–Cornell Institute at Cornell Tech*
                Francesca Rossi, *IBM Research AI*
                Peter Stone, *University of Texas at Austin*
Series ISSN
Print 1939-4608   Electronic 1939-4616

Network Embedding

Theories, Methods, and Applications

Cheng Yang
Beijing University of Posts and Telecommunications, China

Zhiyuan Liu
Tsinghua University, China

Cunchao Tu
Tsinghua University, China

Chuan Shi
Beijing University of Posts and Telecommunications, China

Maosong Sun
Tsinghua University, China

SYNTHESIS LECTURES ON ARTIFICIAL INTELLIGENCE AND MACHINE LEARNING #48

ABSTRACT

Many machine learning algorithms require real-valued feature vectors of data instances as inputs. By projecting data into vector spaces, representation learning techniques have achieved promising performance in many areas such as computer vision and natural language processing. There is also a need to learn representations for discrete relational data, namely networks or graphs. Network Embedding (NE) aims at learning vector representations for each node or vertex in a network to encode the topologic structure. Due to its convincing performance and efficiency, NE has been widely applied in many network applications such as node classification and link prediction.

This book provides a comprehensive introduction to the basic concepts, models, and applications of network representation learning (NRL). The book starts with an introduction to the background and rising of network embeddings as a general overview for readers. Then it introduces the development of NE techniques by presenting several representative methods on general graphs, as well as a unified NE framework based on matrix factorization. Afterward, it presents the variants of NE with additional information: NE for graphs with node attributes/contents/labels; and the variants with different characteristics: NE for community-structured/large-scale/heterogeneous graphs. Further, the book introduces different applications of NE such as recommendation and information diffusion prediction. Finally, the book concludes the methods and applications and looks forward to the future directions.

KEYWORDS

network embedding, network representation learning, node embedding, graph neural network, graph convolutional network, social network, network analysis, deep learning

Contents

Preface

Representation learning techniques, which can extract useful information from data for learning classifiers and other predictors, have achieved promising performance in many areas such as computer vision and natural language processing. There is also a need to learn representations for discrete relational data, namely networks or graphs. By mapping similar nodes in a network to the neighboring regions in the vector space, Network Embedding (NE) is able to learn the latent features that encode the properties of a network. NE has shed a light on the analysis of graph-structured data, and became a hot topic across machine learning and data mining areas during the last five years.

This book provides a comprehensive introduction to the basic concepts, models, and applications of network representation learning (NRL).

We start the book from NE on general graphs where only the topology structure of a network is given for representation learning in Part I. By introducing the history and several representative methods of network embedding, we further propose a general framework for NRL from a theoretic perspective. The framework can cover a number of typical NE algorithms and motivates us to develop an efficient and effective algorithm that can be applied on any NE methods to enhance their performances.

Next, we will introduce NE on graphs with additional information in Part II. In the real world, a vertex in a network usually has rich information, such as text features and other meta data. A joint learning from network structure and these additional information will significantly benefit the quality of network embeddings. We will present NE algorithms with node attributes (Chapter 3), contents (Chapter 5), and labels (Chapter 6) as instances in this part, respectively. We also revisit attributed network embedding from a Graph Convolutional Network(GCN)-based perspective in Chapter 4.

Afterward, we will present network embedding on graphs with different characteristics in Part III, including community-structured graphs (Chapter 7), large-scale graphs (Chapter 8), and heterogeneous graphs (Chapter 9). Community structures widely exist in social networks, and can provide complementary global knowledge to network embeddings which focus on local neighborhoods. Real-world networks are also large-scale and heterogeneous, which motivates us to specialize network embedding algorithms for these kinds of graphs to achieve better efficiency and effectiveness.

We will then show the applications of NE technique on various scenarios in Part IV, namely social relation extraction (Chapter 10), recommendation system (Chapter 11), and information diffusion prediction (Chapter 12). NE respectively serves as the main body/a key part/auxiliary inputs of the algorithms in Chapters 10, 11, and 12.

Finally, we will outlook the future directions of NE and conclude the book in Part V.

The intended readers of this book are researchers and engineers in the fields of machine learning and data mining, especially network analysis. Though the contents in each chapter are mainly self-contained, it is necessary for the readers to have a basic background in machine learning, graph theory, and popular deep learning architectures (e.g., convolutional neural network, recurrent neural network, and attention mechanism). The readers are suggested to quickly understand the area of network embedding through Part I and deeply study the variants in different scenarios through Parts II–IV. We hope the outlook chapter in Part V inspires readers to develop their own NE methods.

Cheng Yang, Zhiyuan Liu, Cunchao Tu, Chuan Shi, and Maosong Sun
February 2021

Acknowledgments

We would like to express our sincere thanks to all those who worked with us on this book. Some parts of this book are based on our previous published or pre-printed papers [Cui et al., 2020, Lu et al., 2019, Tu et al., 2016a,b, 2017a,b, Yang et al., 2015, 2017a,b, 2018a, Zhang et al., 2018b], and we are grateful to our co-authors in the work of network embedding: Edward Chang, Ganqu Cui, Zhichong Fang, Shiyi Han, Linmei Hu, Leyu Lin, Han Liu, Haoran Liu, Yuanfu Lu, Huanbo Luan, Hao Wang, Xiangkai Zeng, Deli Zhao, Wayne Xin Zhao, Bo Zhang, Weicheng Zhang, Zhengyan Zhang, and Jie Zhou. In addition, this work is supported by the National Key Research and Development Program of China (No. 2018YFB1004503), the National Natural Science Foundation of China (No. 62002029, U20B2045, 61772082), and the Fundamental Research Funds for the Central Universities (No. 2020RC23). We would also thank all the editors, reviewers, and staff who helped with the publication of the book. Finally, we would like to thank our families for their wholehearted support.

Cheng Yang, Zhiyuan Liu, Cunchao Tu, Chuan Shi, and Maosong Sun
February 2021

PART I

Introduction to Network Embedding

CHAPTER 1

The Basics of Network Embedding

Machine learning algorithms based on data representation learning have experienced great success in the past few years. Representations learning of data can extract useful information for learning classifiers and other predictors. Distributed representation learning has been widely used in many machine learning tasks, such as computer vision and natural language processing. During the last decade, many works have also been proposed for NE learning and achieve promising results in many important applications. In this chapter, we will start by introducing the background and motivation of NE. Then we will present the origin of NE and formalize the problem. Finally, we will give an overview of this book.

1.1 BACKGROUND

Network (or graph), which is made up of a set of nodes (or vertices) and the connections between them, is an essential data type widely used in our daily lives and academic researches, such as friendship networks in Facebook and citation networks in DBLP [Ley, 2002]. Researchers have extensively studied on many important machine learning applications in networks, such as node classification [Sen et al., 2008], community detection [Yang and Leskovec, 2013], link prediction [Liben-Nowell and Kleinberg, 2007], and anomaly detection [Bhuyan et al., 2014]. Most supervised machine learning algorithms applied on these applications require a set of informative numerical features as input [Grover and Leskovec, 2016]. Therefore, how to numerically represent a network is a key problem for network analysis.

Different from vision and text, graph-structured data is rather "global" and cannot be separated into independent small samples such as images and documents. A typical and straightforward representation of a network is the adjacency matrix, which is a square matrix whose dimension is equal to the number of nodes in the network. The element in the i-th row and j-th column of the adjacency matrix indicates whether there is a directed edge between the i-th and j-th nodes. Theoretically, the number of non-zero elements in the adjacency matrix, i.e., the number of edges (# edges) in the network, may go up to the square of the number of nodes (# nodes). However, a real-world network is usually sparse and # edges is assumed to be linear to # nodes. For example, each user in a social network will link to only a small number of friends, which can be treated as a small constant compared with all users in the network. Consequently,

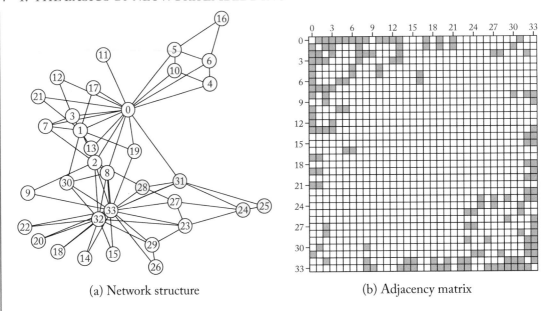

(a) Network structure   (b) Adjacency matrix

Figure 1.1: An illustrative example of a network and its corresponding adjacency matrix. White blocks indicate zero entries.

the adjacency matrix is very sparse and most of its elements are zeros. We present an example of the famous Karate club social network in Fig. 1.1.

Though intuitive and easy to understand, the representation of adjacency matrix has two major disadvantages: high dimensionality and data sparsity. High dimensionality means that each node needs a vector whose length is equal to # nodes to represent, which increases the amount of computation in the subsequent steps. Data sparsity refers to the fact that non-zero elements in the matrix are very sparse, and thus the amount of information encoded in the representations is limited. These two shortcomings make it infeasible to apply machine learning and deep learning techniques to the traditional adjacency matrix representation.

Hand-crafted features may suit the need. For example, we can extract a set of features for each node: node degree, PageRank score, betweenness coefficient, etc. However, such feature engineering requires much human effort and expert knowledge. Also, the extracted features cannot be well generalized to different datasets. To conclude, representing networks numerically is an important prerequisite for network analysis. There are still several challenges unsolved in conventional representation methods.

1.2   THE RISING OF NETWORK EMBEDDING

Representation learning [Bengio et al., 2013] which learns feature embeddings via optimization learning was proposed to avoid feature engineering and improve the flexibility of features. Inspired by recent trends of representation learning on image, speech and natural language, NRL, or NE is proposed as an efficient technique to address the mentioned challenges. NE aims to encode the structure information of each vertex into a low-dimensional real-valued vector, which can be further utilized as features in various network analysis tasks.

We formalize the problem of network representation learning as follows. Given a network $G = (V, E)$ where V is the vertex set and E is the edge set, we want to build a low-dimensional representation $r_v \in \mathbb{R}^k$ for each vertex $v \in V$, where k is expected to be much smaller than $|V|$.

As a dense real-valued representation, r_v can alleviate the sparsity of network representations such as adjacency matrix. We can regard r_v as features of vertex v and apply the features to many machine learning tasks like node classification. The features can be conveniently fed to many classifiers, e.g., logistic regression and Support Vector Machine (SVM). Also, note that the representation is not task-specific and can be shared among different tasks.

NRL facilitates us to better understand the semantic relatedness between vertices, and further alleviates the inconveniences caused by sparsity. Figure 1.2 presents the 2-dimensional network embedding of the Karate graph in Fig. 1.1 learned by DeepWalk [Perozzi et al., 2014]. From Fig. 1.2, we can see that nodes with similar structural roles in the network have similar embeddings in the vector space. In other words, the structural information is preserved in the learned node representations. NE provides an efficient and effective way to represent and manage large-scale networks, alleviating the computation and sparsity issues of conventional symbol-based representations. Hence, NE has attracted much attention in recent years [Grover and Leskovec, 2016, Perozzi et al., 2014, Tang et al., 2015b],[1] and achieves promising performance on many network analysis tasks including link prediction, node classification, and community detection.

1.3   THE EVALUATION OF NETWORK EMBEDDING

The quality of learned NEs are hard to be evaluated directly. Therefore, researchers usually apply the representations on various downstream tasks, and use the performance of downstream tasks as the evaluation of NEs. The most commonly used evaluation tasks are node classification, link prediction, and node clustering (or community detection).

---

[1]In fact, there are some pioneering work [Belkin and Niyogi, 2001, Chen et al., 2007, Tang and Liu, 2009] proposed to leverage structural information into low-dimensional embeddings before 2010. We will give a more detailed introduction about these methods in next chapter.

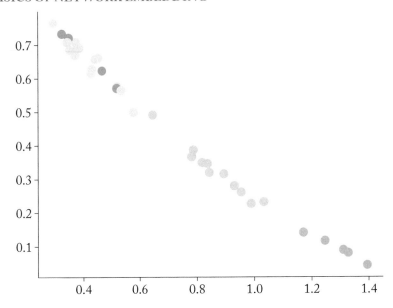

Figure 1.2: An illustration of a typical network embedding method DeepWalk [Perozzi et al., 2014]. Taking the network structure in Fig. 1.1 as input, DeepWalk learns a 2-dimensional vector for each node in the graph.

1.3.1   NODE CLASSIFICATION

Given graph $G = (V, E)$ with a subset of nodes $V_L \subset V$ labeled, where V is the vertex set and E is the edge set, node classification targets on predicting the node labels for every node v in unlabeled node set $V_U \subset V$.

For *unsupervised* NE methods, a standard post-processing classifier such as logistic regression or SVM will be trained by taking learned node representations as input features. Note that the network structure is only used for learning representations and will not be utilized for training classifiers.

For *semi-supervised* NE methods, the class labels of nodes in V_L are utilized during the training of node representations. These methods usually contain a classification module in their models and thus do not require a post-processing classifier. We will introduce more details about semi-supervised methods in Chapter 6.

Node classification problem can be further categorized into multi-class and multi-label classifications.

For *multi-class* classification, each node has exact one label from the whole label set and classification accuracy is usually used as the evaluation metric. For *multi-label* classification, each node may have multiple labels and we will make *number of nodes* × *number of labels* predictions in total. In this case, Micro-F1 (F1 measure of all predictions) and Macro-F1 (average F1 measure over all classes) are usually employed as evaluation metrics.

1.3.2   LINK PREDICTION

Given a partially observed graph $G = (V, E)$, the goal of link prediction is to find the missing links in the network. Formally, a link prediction method will quantify the similarity score of every node pair $(v_i, v_j) \notin E$. The scores of missing links are expected to be better than those of non-existent links.

To adopt NE methods for link prediction, we usually compute the dot product or L1/L2-norm distances between corresponding node representations as the similarity score. For dot product function, a higher score indicates two nodes have more affinities. While the lower score indicates two nodes are more similar for L1-norm and L2-norm functions.

Link prediction is often regarded as an information retrieval problem: missing links are the correct "documents" to retrieve and thus should have top rankings in terms of similarity scores. Standard link prediction metrics include AUC [Hanley and McNeil, 1982] and MRR [Voorhees et al., 1999]. Given the similarity of all vertex pairs, Area Under Curve (AUC) is the probability that a random unobserved link has higher similarity than a random nonexistent link. The mean reciprocal rank (MRR) is a statistical measure for evaluating the ranks of the scores of unobserved links among random nonexistent links, where all links have the same head node and different tail nodes. In fact, most metrics for information retrieval can be employed for evaluating link prediction as well.

1.3.3   NODE CLUSTERING

Given graph $G = (V, E)$, node clustering aims at partitioning the vertices into K disjoint subsets $V = V_1 \cup V_2 \ldots V_K$ where nodes in the same subset are expected to be more similar (or densely connected) than those from different subsets. The number of subsets K can be either predefined or learned from clustering algorithms.

Similar to node classification, a post-processing algorithm such as Spectral Clustering [Ng et al., 2002] will be performed on the similarity matrices of learned embeddings to obtain the node-cluster affiliations.

The evaluation metrics of node clustering depend on whether we are aware of the ground truth clusters. If we have no idea of ground truth clusters, we can use modularity [Newman, 2006] or the Davies–Bouldin index [Davies and Bouldin, 1979] to evaluate whether the vertices in a cluster are densely connected or closely located. If we already have the ground truth clusters, we need another post-processing algorithm such as Hungarian algorithm [Kuhn, 1955] to generate a mapping between predicted clusters and true clusters. Then we can employ Normalized Mutual Information (NMI) or Adjusted Rand Index (ARI) [Gan et al., 2007] metrics for evaluation.

So far, we have introduced the basics of NE. We will give more detailed introduction to typical network representation learning methods in the next chapter.

CHAPTER 2

Network Embedding for General Graphs

In this chapter, we will introduce the most general and popular kind of NE methods which concentrates on learning node representations from network topology without additional information (e.g., node features) or assumptions (e.g., the existence of community structures). Though network representation learning becomes a hot topic of machine learning in last five years, there has been a long history of NE methods since early 2000s. We will first give an overview of the development of NE techniques by introducing several representative algorithms in chronological order. Afterward, we will propose a general framework for NE based on matrix factorization and show that the typical methods are actually special cases of our framework from the theoretic perspective. Finally, with the help of our framework, we propose an efficient and effective algorithm which can be applied on any NE methods to enhance their performances.

2.1 REPRESENTATIVE METHODS

2.1.1 EARLY WORK (EARLY 2000s – EARLY 2010s)

There has been a long history of NE methods since early 2000s. Early works [Roweis and Saul, 2000, Tenenbaum et al., 2000] which were originally proposed for dimension reduction built a neighborhood graph of data points and then leveraged the information in the graph into low-dimensional vectors. Their optimization objectives assumed that the representations of connected nodes should have closer distances and were eventually converted to a solvable eigenvector computation problem. For example, Laplacian Eigenmap [Belkin and Niyogi, 2001] employed the square of Euclidean distance to measure the similarity of node representations, and proposed to minimize the sum of the distances of all connected nodes:

$$\min_{R} \sum_{(v_i, v_j) \in E} \|r_i - r_j\|^2, \qquad (2.1)$$

where R is a $|V|$-by-d matrix where the i-th row of R is a d-dimensional embedding r_i of node v_i. To avoid the trivial all zero solution, Laplacian Eigenmap also added a constraint:

$$R^T D R = I_d, \qquad (2.2)$$

where D is the $|V|$-by-$|V|$ degree matrix where each entry D_{ii} is the degree of node v_i and I_d is the d-by-d indentity matrix.

Table 2.1: Comparisons between DeepWalk and word2vec

Method	Object	Input	Output
word2vec	word	sentence	word embedding
DeepWalk	node	random walk	node embedding

The Laplacian matrix L of a graph is defined as the difference of diagonal matrix D and adjacency matrix A, i.e., $L = D - A$. Thus, the optimization objective can be rewritten as the form of matrix trace

$$\min_{R} \operatorname{tr}(R^T L R). \tag{2.3}$$

It has been proved that the optimal solution R^* of Eq. (2.3), which also satisfies Eq. (2.2), is the eigenvectors of d smallest nonzero eigenvalues of Laplacian matrix L.

By introducing Laplacian Eigenmap as an example, so far we have presented the framework of early methods: (1) set up an optimization objective; (2) reformalization in matrix form; and (3) prove the optimal solution can be obtained by eigenvector computation. This framework was widely used for about 10 years until the early 2010s. The follow-up work paid more attention to the properties of a network. For instance, Directed Graph Embedding (DGE) [Chen et al., 2007] focused on designing asymmetric objective for directed graph and SocioDim [Tang and Liu, 2011] incorporated modularity, which measures how far a network is away from a uniform random one, into the optimization objective. However, the time and space complexities of eigenvector computation could go up to $O(|V|^2)$ and prevent this line of work scaling to large networks.

2.1.2 MODERN WORK (AFTER 2014)

With the great success of representation learning techniques in natural language processing area, DeepWalk [Perozzi et al., 2014] was proposed in 2014 by adopting word2vec [Mikolov et al., 2013b], a well-known word embedding model, for learning node representations. Table 2.1 shows the relationship between DeepWalk and Word2vec. First, DeepWalk made an analogy between node→word and sentence→random walk by showing that node frequency in short random walks also follows power law, just like word frequency in documents. Then DeepWalk applied word2vec with Skip-Gram and hierarchical softmax model on sampled random walks for node representation learning. Now we will give a formal introduction to DeepWalk which plays an important role in network embedding learning area.

Formally, given graph $G = (V, E)$, a random walk $(v_1, v_2, \ldots, v_i)$ is a node sequence started from node v_1 and each node v_k is generated randomly from node v_{k-1}'s neighbors. It is feasible to leverage the topology structure information into random walks. In fact, random walk sequences have been used in many network analysis tasks, such as similarity measurement [Fouss et al., 2007] and community detection [Andersen et al., 2006].

Similar to language modeling, DeepWalk proposed to model short random walk sequences by estimating the likelihood of observing node v_i given all previous ones in the random walk:

$$P(v_i|(v_1, v_2, \ldots, v_{i-1})). \tag{2.4}$$

A simplification of Eq. (2.4) in language modeling is to use node v_i to predict its neighboring nodes $v_{i-w}, \ldots, v_{i-1}, v_{i+1}, \ldots, v_{i+w}$ in the random walk sequence instead, where w is the window size. The neighboring nodes are called as context nodes of the central one and this model is named as Skip-Gram in word representation learning. Therefore, the optimization function of a single node v_i in a random walk can be formulated as

$$\min - \log P((v_{i-w}, \ldots, v_{i-1}, v_{i+1}, \ldots, v_{i+w})|v_i). \tag{2.5}$$

Based on the independent assumption, DeepWalk further simplified the above formula by ignoring the order and offset of the nodes. The loss function can be rewritten as

$$\min \sum_{k=-w, k\neq 0}^{w} - \log P(v_{i+k}|v_i). \tag{2.6}$$

The overall loss function can be obtained by adding up over every node in every sampled random walk.

Now the last piece of DeepWalk is to model the probability $P(v_j|v_i)$ of a context-node pair v_j and v_i. In DeepWalk, each node v_i has two representations with the same dimension: node representation $r_i \in \mathbb{R}^d$ and context representation $c_i \in \mathbb{R}^d$. The probability $P(v_j|v_i)$ is defined by a softmax function over all nodes:

$$P(v_j|v_i) = \frac{\exp(r_i \cdot c_j)}{\sum_{k=1}^{|V|} \exp(r_i \cdot c_k)}, \tag{2.7}$$

where $\cdot$ represents the inner product operation.

Finally, DeepWalk employed hierarchical softmax as an approximation of softmax function for efficiency and utilized stochastic gradient descent method for parameter learning.

DeepWalk outperforms traditional network embedding methods on both efficiency and effectiveness. There are two major advantages of DeepWalk: (1) As a shallow neural network model, DeepWalk employed stochastic gradient descent (SGD) for parameter training instead of transforming the problem into eigenvector computation, which greatly accelerates the training process and makes DeepWalk scalable for large-scale networks. (2) DeepWalk utilized random walks instead of adjacency matrix to characterize the network structure. Compared with adjacency matrix, random walks can further capture the similarity between disconnected nodes, because two nodes sharing many common neighbors will probably appear as a node-context pair in a random walk. Thus, random walks can provide more information and achieve better performance on downstream tasks. In addition, the training process of each random walk only requires local information, which makes streaming/online/distributed algorithms feasible to further improve the time and space efficiency of DeepWalk.

NE has become a rising topic in machine learning and data mining areas since DeepWalk was proposed in 2014.

LINE [Tang et al., 2015b] models first-order and second-order proximities between vertices for learning large-scale NEs. Here the first-order proximity denotes nodes with observed links and second-order proximity represents nodes with shared neighbors.

In specific, LINE parameterizes the first-order proximity between node v_i and v_j as the probability

$$p_1(v_i, v_j) = \frac{1}{1 + \exp(-r_i \cdot r_j)}, \tag{2.8}$$

where r_i and r_j are the embeddings of corresponding nodes.

The target probability is defined as the weighted average $\hat{p}_1(v_i, v_j) = w_{ij} / \sum_{(v_i, v_j) \in E} w_{ij}$ where w_{ij} is the edge weight. Thus, the training objective is to minimize the distance between probability p_1 and $\hat{p}_1$:

$$\mathcal{L}_1 = D_{\mathrm{KL}}(\hat{p}_1 \,\|\, p_1), \tag{2.9}$$

where $D_{\mathrm{KL}}(\cdot \,\|\, \cdot)$ is the KL-divergence of two probability distributions.

On the other hand, the probability that node v_j appears in v_i's context (i.e., v_j is a neighbor of v_i) is parameterized as:

$$p_2(v_j | v_i) = \frac{\exp(c_j \cdot r_i)}{\sum_{k=1}^{|V|} \exp(c_k \cdot r_i)}, \tag{2.10}$$

where c_j is the context embedding of node v_j. Note that if two nodes share many common neighbors, their embeddings will have large inner products with the context embeddings of their common neighbors. Therefore, the embeddings of such nodes will be similar and can capture the second-order proximity.

Similarly, the target probability is defined as $\hat{p}_2(v_j | v_i) = w_{ij} / \sum_k w_{ik}$ and the training objective is to minimize

$$\mathcal{L}_2 = \sum_i \sum_k w_{ik} D_{\mathrm{KL}}(\hat{p}_2(\cdot, v_i) \,\|\, p_2(\cdot, v_i)). \tag{2.11}$$

Finally, the first-order and second-order proximity embeddings are trained independently and concatenated after the training phase as node representations. LINE can be applied on (un)directed or weighted graphs for NE learning.

Node2vec [Grover and Leskovec, 2016] further generalizes DeepWalk with Breadth-First Search (BFS) and Depth-First Search (DFS) on random walks: DeepWalk generates rooted random walks by choosing the next node from a uniform distribution, while node2vec designs a neighborhood sampling strategy which can smoothly interpolate between BFS and DFS. The intuition behind is that BFS can capture local neighborhoods from a microscopic view, and DFS can represent a macro-view of the neighborhood which could encode the community information.

More specifically, assume that we have a random walk which arrives at node v through edge (t, v), node2vec defines the unnormalized transition probability of edge (v, x) for next walk step as $\pi_{vx} = \alpha_{pq}(t, x) \cdot w_{vx}$, where

$$\alpha_{pq}(t, x) = \begin{cases} \frac{1}{p} & \text{if} \quad d_{tx} = 0 \\ 1 & \text{if} \quad d_{tx} = 1 \\ \frac{1}{q} & \text{if} \quad d_{tx} = 2 \end{cases} \qquad (2.12)$$

and d_{tx} denotes the shortest path distance between nodes t and x. p and q are hyper-parameters controlling the behaviors of random walks. A small p will increase the probability of revisiting a node and limit the random walk in a local neighborhood while a small q will encourage the random walk to explore distant nodes.

GraRep [Cao et al., 2015] factorizes different k-order proximity matrices and concatenates the embeddings learned from each proximity matrix. SDNE [Wang et al., 2016a] employs deep neural model for learning embeddings. Here we only list several most influential follow-up work and more algorithms will be detailed in later chapters. In the next section, we will introduce a general framework for NE.

2.2   THEORY: A UNIFIED NETWORK EMBEDDING FRAMEWORK

In this section, we summarize several representative NE methods into a unified two-step framework, including proximity matrix construction and dimension reduction. The first step builds a proximity matrix M where each entry M_{ij} encodes the proximity information between vertex i and j. The second step reduces the dimension of the proximity matrix to obtain NEs. Different NE methods employ various dimension reduction algorithms such as eigenvector computation and SVD decomposition. Our analysis of the first step, i.e., proximity matrix construction, shows that the quality of NEs can be improved when higher-order proximities are encoded into the proximity matrix.

2.2.1   K-ORDER PROXIMITY

First, we clarify the notations and introduce the concept of k-order proximity. Let $G = (V, E)$ be a given network where V is node set and E is edge set. The task is to learn a real-valued representation $r_v \in \mathbb{R}^d$ for each vertex $v \in V$ where d is the embedding dimension. We assume that networks are unweighted and undirected in this chapter without loss of generality and define adjacency matrix $\widetilde{A} \in \mathbb{R}^{|V| \times |V|}$ as $\widetilde{A}_{ij} = 1$ if $(v_i, v_j) \in E$ and $\widetilde{A}_{ij} = 0$ otherwise. Diagonal matrix $D \in \mathbb{R}^{|V| \times |V|}$ denotes the degree matrix where $D_{ii} = d_i$ represents the degree of vertex

v_i. $A = D^{-1}\widetilde{A}$ is normalized adjacency matrix where the sum of each row equals to 1. Similarly, we also have Laplacian matrix $\widetilde{L} = D - \widetilde{A}$ and normalized Laplacian matrix $L = D^{-\frac{1}{2}}\widetilde{L}D^{-\frac{1}{2}}$.

The (normalized) adjacency matrix and Laplacian matrix characterize the first-order proximity which models the local pairwise proximity between vertices. Note that each off-diagonal nonzero entry of the first-order proximity matrix corresponds to an edge in the network. However, real-world networks are always sparse which indicates that $O(E) = O(V)$. Therefore, the first-order proximity matrix is usually very sparse and insufficient to fully model the pairwise proximity between vertices. As a result, people also explore higher-order proximity to model the strength between vertices [Cao et al., 2015, Perozzi et al., 2014, Tang et al., 2015b]. For example, the second-order proximity can be characterized by the number of common neighbors between vertices. As an alternative view, the second-order proximity between v_i and v_j can also be modeled by the probability that a two-step random walk from v_i reaches v_j. Intuitively, the probability will be large if v_i and v_j share many common neighbors. In the probabilistic setting based on random walk, we can easily generalize it to k-order proximity [Cao et al., 2015]: the probability that a random walk starts from v_i and walks to v_j with exactly k steps. Note that the normalized adjacency matrix A is the transition probability matrix of a single-step random walk. Then we can compute k-step transition probability matrix as the k-order proximity matrix

$$A^k = \underbrace{A \cdot A \ldots A}_{k}, \tag{2.13}$$

where the entry A_{ij}^k is the k-order proximity between node v_i and v_j.

2.2.2 NETWORK EMBEDDING FRAMEWORK

Now we have introduced the concept of k-order proximity matrix. In this subsection, we will present the framework based on dimension reduction of a proximity matrix and conduct a theoretical analysis to show that several aforementioned NE methods including Social Dimension [Tang and Liu, 2011], DeepWalk [Perozzi et al., 2014], GraRep [Cao et al., 2015], TADW [Yang et al., 2015], and LINE [Tang et al., 2015b] can be formalized into this framework.

We summarize NE methods as a two-step framework.

Step 1: Proximity Matrix Construction. Compute a proximity matrix $M \in \mathbb{R}^{|V| \times |V|}$ which encodes the information of k-order proximity matrix where $k = 1, 2 \ldots, K$. For example, $M = \frac{1}{K}A + \frac{1}{K}A^2 \cdots + \frac{1}{K}A^K$ stands for an average combination of k-order proximity matrix for $k = 1, 2 \ldots, K$. The proximity matrix M is usually represented by a polynomial of normalized adjacency matrix A of degree K and we denote the polynomial as $f(A) \in \mathbb{R}^{|V| \times |V|}$. Here the degree K of polynomial $f(A)$ corresponds to the maximum order of proximities encoded in proximity matrix. Note that the storage and computation of proximity matrix M does not necessarily take $O(|V|^2)$ time because we only need to save and compute the nonzero entries.

Step 2: Dimension Reduction. Find NE matrix $R \in \mathbb{R}^{|V| \times d}$ and context embedding $C \in \mathbb{R}^{|V| \times d}$ so that the product $R \cdot C^T$ approximates proximity matrix M. Here different algorithms may employ different distance functions to minimize the distance between M and $R \cdot C^T$. For example, we can naturally use the norm of matrix $M - R \cdot C^T$ to measure the distance and minimize it.

Now we will show how popular algorithms can be fit into the proposed framework.

Social Dimension [Tang and Liu, 2011] computes the first d eigenvectors of normalized Laplacian matrix L as d-dimensional network representations. The information embedded in the eigenvectors comes from the first-order proximity matrix L. Note that the real-valued symmetric matrix L can be factorized as $L = Q \Lambda Q^{-1}$ via eigendecomposition[1] where $\Lambda \in \mathbb{R}^{|V| \times |V|}$ is a diagonal matrix, $\Lambda_{11} \geq \Lambda_{22} \geq \ldots \Lambda_{|V||V|}$ are eigenvalues and $Q \in \mathbb{R}^{|V| \times |V|}$ is the eigenvector matrix.

We can equivalently transform Social Dimension into our framework by setting proximity matrix M as the first-order proximity matrix L, NE R as the first d columns of eigenvector matrix Q and context embedding C^T as the first d rows of ΛQ^{-1}.

DeepWalk [Perozzi et al., 2014] generates random walks and employs Skip-Gram model for representation learning. DeepWalk learns two representations for each vertex and we denote the NE and context embedding as matrix $R \in \mathbb{R}^{|V| \times d}$ and $C \in \mathbb{R}^{|V| \times d}$. Our goal is to figure out the closed form of matrix M where $M = RC^T$.

Suppose that we have the node-context set D generated from random walk sequences, where each member of D is a node-context pair (v, c). For each node-context pair (v, c), $N(v, c)$ denotes how many times that (v, c) appears in D. $N(v) = \sum_{c' \in V_C} N(v, c')$ and $N(c) = \sum_{v' \in V} N(v', c)$ denote how many times that v and c appear in D, respectively.

It has been shown that Skip-Gram with Negative Sampling (SGNS) is implicitly factorizing a word-context matrix M [Levy and Goldberg, 2014] by assuming that dimensionality d is sufficiently large. Each entry in M is

$$M_{ij} = \log \frac{N(v_i, c_j) \cdot |D|}{N(v_i) \cdot N(c_j)} - \log n, \tag{2.14}$$

where n is the number of negative samples for each word-context pair. M_{ij} can be interpreted as Pointwise Mutual Information (PMI) of word-context pair (v_i, c_j) shifted by $\log n$. Similarly,[2] we can prove that Skip-Gram with softmax is factorizing a matrix M where

$$M_{ij} = \log \frac{N(v_i, c_j)}{N(v_i)}. \tag{2.15}$$

Now we discuss what M_{ij} in DeepWalk represents. It is clear that the method of sampling node-context pairs will affect matrix M. Assume that the network is connected and undirected,

---

[1]https://en.wikipedia.org/wiki/Eigendecomposition_of_a_matrix
[2]The detailed proof can be found at our note [Yang and Liu, 2015] on arxiv.

we will discuss $N(v)/|D|$, $N(c)/|D|$ and $N(v, c)/N(v)$ based on an ideal sampling method for DeepWalk algorithm: first we generate a sufficiently long random walk RW. Denote RW_i as the node on position i of RW. Then we add node-context pair (RW_i, RW_j) into D if and only if $0 < |i - j| \leq w$ where w is the window size used in Skip-Gram model.

Each occurrence of node i will be recorded $2w$ times in D for undirected graph. Thus, $N(v_i)/|D|$ is the frequency of v_i appears in the random walk, which is exactly the PageRank value of v_i. Also, note that $2wN(v_i, v_j)/N(v_i)$ is the expectation times that v_j is observed in left or right w neighbors of v_i. Now we try to figure out $N(v_i, v_j)/N(v_i)$ based on this comprehension.

The transition matrix in PageRank is exactly the normalized adjacency matrix A. We use e_i to denote a $|V|$-dimensional row vector where all entries are 0 except the i-th entry is 1. Suppose that we start a random walk from vertex i and use e_i to denote the initial state. Then $e_i A$ is the distribution over all the nodes and the j-th entry of $e_i A$ is the probability that node i walks to node j. Hence, the j-th entry of $e_i A^w$ is the probability that node i walks to node j at exactly w steps. Thus, $[e_i(A + A^2 + \cdots + A^w)]_j$ is the expectation times that v_j appears in right w neighbors of v_i. Hence, we have

$$\frac{N(v_i, v_j)}{N(v_i)} = \frac{[e_i(A + A^2 + \cdots + A^w)]_j}{w}. \tag{2.16}$$

This equality also holds for directed graph. Hence, we can see that $M_{ij} = \log N(v_i, v_j)/N(v_i)$ is the logarithm of the average probability that node i randomly walks to node j in w steps.

To conclude, DeepWalk implicitly factorizes a matrix $M \in \mathbb{R}^{|V| \times |V|}$ into the product of $R \cdot C^T$, where

$$M = \log \frac{A + A^2 + \cdots + A^w}{w}, \tag{2.17}$$

and w is the window size used in Skip-Gram model. Matrix M characterizes the average of the first-order, second-order, $\ldots$, w-order proximities. DeepWalk algorithm approximates high-order proximity by Monte Carlo sampling based on random walk generation without calculating the k-order proximity matrix directly.

To adopt DeepWalk algorithm to our two-step framework, we can simply set proximity matrix $M = f(A) = \frac{A + A^2 + \cdots + A^w}{w}$. Note that we omit the log operation in Eq. (2.17) to avoid numerical issues.

GraRep [Cao et al., 2015] accurately calculates k-order proximity matrix A^k for $k = 1, 2 \ldots, K$, computes a specific representation for each k, and concatenates these embeddings. Specifically, GraRep reduces the dimension of k-order proximity matrix A^k for k-order representation via SVD decomposition. In detail, we assume that k-order proximity matrix A^k is factorized into the product of $U \Sigma S$ where $\Sigma \in \mathbb{R}^{|V| \times |V|}$ is a diagonal matrix, $\Sigma_{11} \geq \Sigma_{22} \geq \ldots \Sigma_{|V||V|} \geq 0$ are singular values and $U, S \in \mathbb{R}^{|V| \times |V|}$ are unitary matrices. GraRep defines k-order NE and context embedding $R_{\{k\}}, C_{\{k\}} \in \mathbb{R}^{|V| \times d}$ as the first d columns of $U \Sigma^{\frac{1}{2}}$ and $S^T \Sigma^{\frac{1}{2}}$,

Table 2.2: Comparisons among three NE methods

	SD	DeepWalk	GraRep
Proximity Matrix	L	$\sum_{k=1}^{K} \frac{A^k}{K}$	$A^k, k = 1 \ldots K$
Computation	Accurate	Approximate	Accurate
Scalable	Yes	Yes	No
Performance	Low	Middle	High

respectively. The computation of k-order representation $R_{\{k\}}$ naturally follows our framework. However, GraRep cannot efficiently scale to large networks [Grover and Leskovec, 2016]: though the first-order proximity matrix A is sparse, a direct computation of A^k ($k \geq 2$) takes $O(|V|^2)$ time which is unacceptable for large-scale networks.

Text-Associated DeepWalk (TADW) [Yang et al., 2015] and LINE [Tang et al., 2015b] can also be formalized into our framework in a similar way.

2.2.3   OBSERVATIONS

By far, we have shown five representative NE algorithms can be formulated into our two-step framework, i.e., proximity matrix construction and dimension reduction. In this subsection, we focus on the first step and study how to define a good proximity matrix for NE. The study of different dimension reduction methods, e.g., SVD decomposition, will be left as future work.

We summarize the comparisons among Social Dimension (SD), DeepWalk, and GraRep in Table 2.2 and conclude the following observations.

Observation 1 Modeling higher order and accurate proximity matrix can improve the quality of network representation. In other words, NE could benefit if we explore a polynomial proximity matrix $f(A)$ of a higher degree.

From the development of NE methods, we can see that DeepWalk outperforms SD because DeepWalk considers higher-order proximity matrices and the higher-order proximity matrices can provide complementary information for lower-order proximity matrices. GraRep outperforms DeepWalk because GraRep accurately calculates the k-order proximity matrix rather than approximating it by Monte Carlo simulation as DeepWalk did.

Observation 2 Accurate computation of high-order proximity matrix is not feasible for large-scale networks.

The major drawback of GraRep is the computation complexity of calculating the accurate k-order proximity matrix. In fact, the computation of high-order proximity matrix takes $O(|V|^2)$ time and the time complexity of SVD decomposition also increases as k-order proximity matrix gets dense when k grows. In summary, a time complexity of $O(|V|^2)$ is too expensive to handle large-scale networks.

The first observation motivates us to explore higher-order proximity matrix in NE algorithm but the second observation prevents us from an accurate inference of higher-order proximity matrices. A possible solution to this dilemma is to study the problem that how to learn NEs from approximate higher-order proximity matrices efficiently. We will formalize the problem and introduce our algorithm in next section.

2.3   METHOD: NETWORK EMBEDDING UPDATE (NEU)

An accurate computation of high-order proximities is time consuming and thus not scalable for large-scale networks. Thus, we can only approximate higher-order proximity matrix for learning better NEs. In order to be more efficient, we also seek to use the network representations which encode the information of lower-order proximities as our basis to avoid repeated computations. Therefore, we propose NE Update (NEU) algorithm which could be applied to any NE methods to enhance their performances. The intuition behind is that the embeddings processed by NEU algorithm can implicitly approximate higher order proximities with a theoretical approximation bound and thus achieve better performances.

We conduct experiments on multi-label classification and link prediction tasks over three publicly available datasets to evaluate the quality of NEs. Experimental results show that the NEs learned by existing methods can be improved consistently and significantly on both evaluation tasks after enhanced by NEU. Moreover, the running time of NEU takes less than 1% training time of popular NE methods such as DeepWalk and LINE, which could be negligible.

2.3.1   PROBLEM FORMALIZATION

In order to be more efficient, we aim to use the network representations which encode the information of lower-order proximity matrices as our basis to avoid repeated computations. We formalize our problem below.

Problem Formalization: Assume that we have normalized adjacency matrix A as the first-order proximity matrix, NE R and context embedding C where $R, C \in \mathbb{R}^{|V| \times d}$. Suppose that the embeddings R and C are learned by the above NE framework which indicates that the product $R \cdot C^T$ approximates a polynomial proximity matrix $f(A)$ of degree K. Our goal is to learn a better representation R' and C' which approximates a polynomial proximity matrix $g(A)$ with higher degree than $f(A)$. Also, the algorithm should be efficient in the linear time of $|V|$. Note that the lower bound of time complexity is $O(|V|d)$ which is the size of embedding matrix R.

2.3.2   APPROXIMATION ALGORITHM

In this subsection, we present a simple, efficient, and effective iterative updating algorithm to solve the above problem.

Method: Given hyperparameter $\lambda \in (0, \frac{1}{2}]$, normalized adjacency matrix A, we update NE R and context embedding C as follows:

$$R' = R + \lambda A \cdot R,$$
$$C' = C + \lambda A^T \cdot C. \tag{2.18}$$

The time complexity of computing $A \cdot R$ and $A^T \cdot C$ is $O(|V|d)$ because matrix A is sparse and has $O(|V|)$ nonzero entries. Thus, the overall time complexity of one iteration of Eq. (2.18) is $O(|V|d)$.

Recall that product of previous embedding R and C approximates polynomial proximity matrix $f(A)$ of degree K. Now we prove that the algorithm can learn better embeddings R' and C' where the product $R' \cdot C'^T$ approximates a polynomial proximity matrix $g(A)$ of degree $K + 2$ bounded by matrix infinite norm.

Theorem: Given network and context embedding R and C, we suppose that the approximation between $R \cdot C^T$ and proximity matrix $M = f(A)$ is bounded by $r = ||f(A) - R \cdot C^T||_\infty$ and $f(\cdot)$ is a polynomial of degree K. Then the product of updated embeddings R' and C' from Eq. (2.18) approximates a polynomial $g(A) = f(A) + 2\lambda A f(A) + \lambda^2 A^2 f(A)$ of degree $K + 2$ with approximation bound $r' = (1 + 2\lambda + \lambda^2)r \le \frac{9}{4}r$.

Proof: Assume that $S = f(A) - RC^T$ and thus $r = ||S||_\infty$.

$$\begin{aligned}
||g(A) - R'C'^T||_\infty &= ||g(A) - (R + \lambda AR)(C^T + \lambda C^T A)||_\infty \\
&= ||g(A) - RC^T - \lambda ARC^T - \lambda RC^T A - \lambda^2 ARC^T A||_\infty \\
&= ||S + \lambda AS + \lambda SA + \lambda^2 ASA||_\infty \\
&\le ||S||_\infty + \lambda ||A||_\infty ||S||_\infty + \lambda ||S||_\infty ||A||_\infty + \lambda^2 ||S||_\infty ||A||_\infty^2 \\
&= r + 2\lambda r + \lambda^2 r,
\end{aligned} \tag{2.19}$$

where the second last equality replaces $g(A)$ and $f(A) - RC^T$ by the definitions of $g(A)$ and S and the last equality uses the fact that $||A||_\infty = \max_i \sum_j |A_{ij}| = 1$ because the summation of each row of A equals to 1.

In our experimental settings, we assume that the weight of lower-order proximities should be larger than higher-order proximities because they are more directly related to the original network. Therefore, given $g(A) = f(A) + 2\lambda A f(A) + \lambda^2 A^2 f(A)$, we have $1 \ge 2\lambda \ge \lambda^2 > 0$ which indicates that $\lambda \in (0, \frac{1}{2}]$. The proof indicates that the updated embedding can implicitly approximate a polynomial $g(A)$ of 2 more degrees within $\frac{9}{4}$ times matrix infinite norm of previous embeddings. QED.

Algorithm 2.1 Network Embedding Update Algorithm

Require: Hyperparameter λ_1, λ_2, and T, network and context embedding $R, C \in \mathbb{R}^{|V| \times d}$
Ensure: Return network and context embedding R, C
  1: **for** iter from 1 to T **do**
  2:    $R \doteq R + \lambda_1 A \cdot R + \lambda_2 A \cdot (A \cdot R)$
  3:    $C \doteq C + \lambda_1 A^T \cdot C + \lambda_2 A^T \cdot (A^T \cdot C)$
  4: **end for**

Algorithm: The update Eq. (2.18) can be further generalized in two directions. First, we can update embeddings R and C according to Eq. (2.20):

$$R' = R + \lambda_1 A \cdot R + \lambda_2 A \cdot (A \cdot R),$$
$$C' = C + \lambda_1 A^T \cdot C + \lambda_2 A^T \cdot (A^T \cdot C). \tag{2.20}$$

The time complexity is still $O(|V|d)$ but Eq. (2.20) can obtain higher proximity matrix approximation than Eq. 2.18 in one iteration. More complex update formulas which explores further higher proximities than Eq. (2.20) can also be applied but we use Eq. (2.20) in our experiments as a cost-effective choice.

Another direction is that the update equation can be processed for T rounds to obtain higher proximity approximation. However, the approximation bound would grow exponentially as the number of rounds T grows and thus the update cannot be done infinitely. Note that the update operation of R and C are completely independent. Therefore, we only need to update NE R. We name our algorithm as NE Update (NEU). NEU avoids an accurate computation of high-order proximity matrix but can yield NEs that actually approximate high-order proximities. Hence, our algorithm can improve the quality of NEs efficiently. Intuitively, Eqs. (2.18) and (2.20) allow the learned embeddings to further propagate to their neighbors. Thus, proximities of longer distance between vertices will be embedded. The pseudocode of NEU is shown in Algorithm 2.1.

2.4   EMPIRICAL ANALYSIS

We evaluate the qualities of NEs on two tasks: multi-label classification and link prediction. We perform our NEU algorithm over the embeddings learned by baseline methods and report both evaluation performance and running time.

2.4.1   DATASETS

We conduct experiments on three publicly available datasets: Cora [Sen et al., 2008], BlogCatalog, and Flickr [Tang and Liu, 2011]. We assume that all three datasets are undirected and unweighted networks.

Cora contains 2,708 machine learning papers drawn from 7 classes and 5,429 citation links between them. Each paper has exactly one class label. Each paper in Cora dataset also has text information denoted by a 1,433 dimensional binary vector indicating the presence of the corresponding words.

BlogCatalog contains 10,312 bloggers and 333,983 social relationships between them. The labels represent the topic interests provided by the bloggers. The network has 39 labels and a blogger may have multiple labels.

Flickr contains 80,513 users from a photo-sharing website and 5,899,882 friendships between them. The labels represent the group membership of users. The network has 195 labels and a user may have multiple labels.

2.4.2   BASELINES AND EXPERIMENTAL SETTINGS

We consider a number of baselines to demonstrate the effectiveness and robustness of NEU algorithm. For all methods and datasets, we set the embedding dimension $d = 128$.

Graph Factorization (GF) simply factorizes the normalized adjacency matrix A via SVD decomposition to reduce dimensions for NEs.

Social Dimension (SD) [Tang and Liu, 2011] computes the first d eigenvectors of normalized Laplacian matrix as d-dimensional embeddings.

DeepWalk [Perozzi et al., 2014] generates random walks and employs Skip-Gram model for representation learning. DeepWalk has three hyperparameters besides embedding dimension d: window size w, random walk length t, and walks per vertex γ. As these hyperparameters increase, the number of training samples and running time will increase. We evaluate three groups of hyperparameters of DeepWalk, i.e., a default setting of the authors' implementation DeepWalk$_{low}$ where $w = 5, t = 40, \gamma = 10$, the setting used in node2vec [Grover and Leskovec, 2016], DeepWalk$_{mid}$ where $w = 10, t = 80, \gamma = 10$ and the setting used in the original paper [Perozzi et al., 2014] DeepWalk$_{high}$ where $w = 10, t = 40, \gamma = 80$.

LINE [Tang et al., 2015b] is a scalable NE algorithm which models first-order and second-order proximities between vertices using two separate network representations LINE$_{1st}$ and LINE$_{2nd}$, respectively. We use default settings for all hyperparameters except the number of total training samples $s = 10^4|V|$ so that LINE has comparable running time against DeepWalk$_{mid}$.

TADW [Yang et al., 2015] incorporates text information into DeepWalk under the framework of matrix factorization. We add this baseline for Cora dataset.

node2vec [Grover and Leskovec, 2016] is a semi-supervised method which generalizes DeepWalk algorithm with BFS and DFS of random walks. We use the same hyperparameter setting used in their paper: $w = 10, t = 80, \gamma = 10$. We employ a grid search over return parameter and in-out parameter $p, q \in \{0.25, 0.5, 1, 2, 4\}$ for semi-supervised training.

GraRep [Cao et al., 2015] accurately calculates k-order proximity matrix A^k for $k = 1, 2 \ldots, K$, computes a specific representation for each k, and concatenates these embeddings. We only use GraRep [Cao et al., 2015] for the smallest dataset Cora due to its inefficiency [Grover and Leskovec, 2016]. We set $K = 5$ and thus GraRep has $128 \times 5 = 640$ dimensions.

Experimental Settings: For SD, DeepWalk, LINE, node2vec, we directly use the implementations provided by their authors. We set the hyperparameters of NEU as follows: $\lambda_1 = 0.5, \lambda_2 = 0.25$ for all three datasets, $T = 3$ for Cora and BlogCatalog and $T = 1$ for Flickr. Here, λ_1, λ_2 are set empirically following the intuition that lower proximity matrix should have a higher weight and T is set as the maximum iteration before the performance on 10% random validation set begins to drop. In fact, we can simply set $T = 1$ if we have no prior knowledge of downstream tasks. The experiments are executed on a single CPU for the ease of running time comparison and the CPU type is Intel Xeon E5-2620 @ 2.0GHz.

2.4.3 MULTI-LABEL CLASSIFICATION

For multi-label classification task, we randomly select a portion of vertices as training set and leave the rest as test set. We treat NEs as vertex features and feed them into a one-vs-rest SVM classifier implemented by LibLinear [Fan et al., 2008] as previous works did [Tang and Liu, 2009, 2011]. We repeat the process for 10 times and report the average Micro-F1 score. Since a vertex of Cora dataset has exactly one label, we report classification accuracy for this dataset. We normalize each dimension of NEs so that the $L2$-norm of each dimension equals to 1 before we feed the embeddings into the classifier as suggested by Ben-Hur and Weston [2010]. We also perform the same normalization before and after NEU. The experimental results are listed in Tables 2.3, 2.4, and 2.5. The numbers in the brackets represent the performances of corresponding methods after processing NEU. "+0.1,""+0.3,""+1," and "+8" in the time column indicate the additional running time of NEU on Cora, BlogCatalog, and Flickr dataset, respectively. As an illustrative example on Cora dataset, NEU takes 0.1 second and improves the classification accuracy from 78.1 to 84.4 for NEs learned by TADW when labeled ratio is 10%. We exclude node2vec on Flickr as node2vec failed to terminate in 24 hours on this dataset. We **bold** the results when NEU achieves more than 10% relative improvement. We conduct 0.05 level paired t-test and mark all entries that fail to reject the null hypothesis with *.

2.4.4 LINK PREDICTION

For the purpose of link prediction, we need to score each pair of vertices given their embeddings. For each pair of NE r_i and r_j, we try three scoring functions, i.e., cosine similarity $\frac{r_i \cdot r_j}{\|r_i\|_2 \|r_j\|_2}$, inner product $r_i \cdot r_j$, and inverse $L2$-distance $1/\|r_i - r_j\|_2$. We use AUC

Table 2.3: Classification results on Cora dataset

% Labeled nodes	% Accuracy			Time (s)
	10%	**50%**	**90%**	
GF	50.8 (**68.0**)	61.8 (**77.0**)	64.8 (77.2)	4 (+0.1)
SC	55.9 (**68.7**)	70.8 (**79.2**)	72.7 (80.0)	1 (+0.1)
DeepWalk$_{low}$	71.3 (76.2)	76.9 (81.6)	78.7 (81.9)	31 (+0.1)
DeepWalk$_{mid}$	68.9 (**76.7**)	76.3 (82.0)	78.8 (84.3)	69 (+0.1)
DeepWalk$_{high}$	68.4 (**76.1**)	74.7 (80.5)	75.4 (81.6)	223 (+0.1)
LINE$_{1st}$	64.8 (70.1)	76.1 (80.9)	78.9 (82.2)	62 (+0.1)
LINE$_{2nd}$	63.3 (**73.3**)	73.4 (80.1)	75.6 (80.3)	67 (+0.1)
node2vec	76.9 (77.5)	81.0 (81.6)	81.4 (81.9)	56 (+0.1)
TADW	78.1 (84.4)	83.1 (86.6)	82.4 (87.7)	2 (+0.1)
GraRep	70.8 (76.9)	78.9 (82.8)	81.8 (84.0)	67 (+0.3)

Table 2.4: Classification results on BlogCatalog dataset

% Labeled nodes	% Micro-F1			Time (s)
	1%	**5%**	**9%**	
GF	17.0 (**19.6**)	22.2 (**25.0**)	23.7 (**26.7**)	19 (+1)
SC	19.4 (20.3)	26.9 (28.1)	29.0 (31.0)	10 (+1)
DeepWalk$_{low}$	24.5 (26.4)	31.0 (33.4)	32.8 (35.1)	100 (+1)
DeepWalk$_{mid}$	24.0 (**27.1**)	31.0 (33.8)	32.8 (35.7)	225 (+1)
DeepWalk$_{high}$	24.9 (26.4)	31.5 (33.7)	33.7 (35.9)	935 (+1)
LINE$_{1st}$	23.1 (24.7)	29.3 (31.6)	31.8 (33.5)	241 (+1)
LINE$_{2nd}$	21.5 (**25.0**)	27.9 (**31.6**)	30.0 (**33.6**)	244 (+1)
node2vec	25.0 (27.0)	31.9 (34.5)	35.1 (37.2)	454 (+1)

value [Hanley and McNeil, 1982] which indicates the probability that the score of an unobserved link is higher than that of a nonexistent link as our evaluation metric and select the scoring function with best performance for each baseline. We remove 20% edges of Cora, 50% of BlogCatalog and Flickr as test set and use the remaining links to train NEs. We also add three commonly used link prediction baselines for comparison: Common Neighbors (CN), Jaccard Index, and Salton Index [Salton and McGill, 1986]. We only report the best performance for DeepWalk∈{DeepWalk$_{low}$, DeepWalk$_{mid}$, DeepWalk$_{high}$}, and LINE∈{LINE$_{1st}$, LINE$_{2nd}$}

Table 2.5: Classification results on Flickr dataset

% Labeled nodes	% Micro-F1			Time (s)
	1%	5%	9%	
GF	21.1 (21.8)	22.0 (23.1)	21.7 (23.4)	241 (+8)
SC	24.1 (**29.2**)	27.5 (**34.1**)	28.3 (**34.7**)	102 (+8)
DeepWalk$_{low}$	28.5 (**31.4**)	30.9 (33.5)	31.3 (33.8)	1,449 (+8)
DeepWalk$_{mid}$	29.5 (31.9)	32.4 (35.1)	33.0 (35.4)	2,282 (+8)
DeepWalk$_{high}$	31.8 (33.1)	36.3 (36.7)	37.3 (37.6)	9,292 (+8)
LINE$_{1st}$	32.0 (32.7)	35.9 (36.4)	36.8 (37.2)	2,664 (+8)
LINE$_{2nd}$	30.0 (31.0)	34.2 (34.4)	35.1 (35.2)	2,740 (+8)

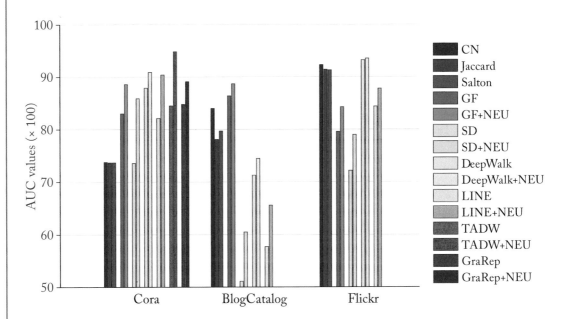

Figure 2.1: Experimental results on link prediction.

and omit the results of node2vec as it only yields comparable and even worse performance than the best performed DeepWalk. The experimental results are shown in Fig. 2.1. For each dataset, the three leftmost columns represent the traditional link prediction baselines. Then each pair of columns stands for a NE method and its performance after NEU.

2.4.5   EXPERIMENTAL RESULTS ANALYSIS

We have four main observations over the experimental results of two evaluation tasks.

(1) NEU consistently and significantly improves the performance of various NEs using almost negligible running time on both evaluation tasks. The absolute improvements on Flickr are not as significant as that on Cora and BlogCatalog because Flickr dataset has an average degree of 147 which is much denser than Cora and BlogCatalog and thus the impact of higher-order proximity information is diluted by rich first-order proximity information. But for Cora dataset where the average degree is 4, NEU has very significant improvement as high-order proximity plays an important role for sparse networks.

(2) NEU facilitates NE method to converge fast and stably. We can see that the performances of DeepWalk$_{low}$+NEU and DeepWalk$_{mid}$+NEU are comparable and even better than that of DeepWalk$_{mid}$ and DeepWalk$_{high}$, respectively, and the former ones use much less time. Also, DeepWalk encounters the overfitting problem on Cora dataset as the classification accuracy drops when hyperparameters increase. However, the performances of DeepWalk+NEU are very stable and robust.

(3) NEU also works for NE algorithms which don't follow our two-step framework, i.e., node2vec. NEU doesn't harm the performance of node2vec and even improve a little bit. This observation demonstrates the effectiveness and robustness of NEU.

(4) NEU can serve as a standard preprocessing step for evaluations of future NE methods. In other words, NEs will be evaluated only after enhanced by NEU as NEU won't increase time and space complexity at all.

2.5   FURTHER READING

In this chapter, we propose a unified NE framework based on matrix factorization. The proposed framework can cover a number of NE methods such as DeepWalk, LINE, and GraRep. There are also some follow-up works [Huang et al., 2018, Liu et al., 2019b, Qiu et al., 2018] discussing the equivalence between NE and matrix factorization. They bring more NE algorithms including PTE [Tang et al., 2015a] and node2vec [Grover and Leskovec, 2016] into the framework of matrix factorization. Note that the equivalent forms of matrix factorization for the same NE method could be different in these works due to different assumptions.

Part of this chapter was published in our IJCAI17 conference paper by Yang et al. [2017a].

PART II

Network Embedding with Additional Information

CHAPTER 3

Network Embedding for Graphs with Node Attributes

Most NE methods investigate network structures for learning. In reality, nodes in a graph usually contain rich feature information, which cannot be well applied with typical representation learning methods. Taking text feature as an example, we will introduce text-associated Deep-Walk (TADW) model for learning NEs with node attributes in this chapter. Inspired by the proof that DeepWalk, a state-of-the-art network representation method, is actually equivalent to matrix factorization (MF), TADW incorporates text features of vertices into network representation learning under the framework of matrix factorization. We evaluate TADW and various baseline methods by applying them to the task of multi-class classification of vertices. The experimental results show that, TADW outperforms other baselines on all three datasets, especially when networks are noisy and training ratio is small.

3.1 OVERVIEW

As shown in Chapter 2, most works in NE learn representations from network structure. For example, social dimensions [Tang and Liu, 2009, 2011] are proposed by computing eigenvectors of Laplacian or modularity matrix of a network. Recently, Skip-Gram, a word representation model in NLP, is introduced to learn node representations from random walk sequences in social networks, dubbed DeepWalk [Perozzi et al., 2014]. Both social dimensions and DeepWalk methods take a network structure as input to learn network representations, without considering any other information.

In the real world, a node in a network usually has rich information, such as text features and other meta data. For example, Wikipedia articles connect to each other and form a network, and each article, as a node, has substantial text information, which may also be important to NE. Hence, we come up with an idea to learn network representations from both network structure and node attributes.

A straightforward method is to learn representations from text features and network features independently, and then concatenate the two separate representations. The method, however, does not take the sophisticated interactions between network structure and text information into consideration, and thus usually leads to no avail. It is also non-trivial to incorporate text information in existing NE frameworks. For example, DeepWalk cannot easily handle additional information during its random walks in a network.

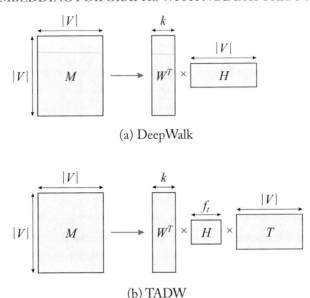

(a) DeepWalk

(b) TADW

Figure 3.1: (a) DeepWalk as matrix factorization. (b) Text-associated matrix factorization (TADW).

Fortunately, given a network $G = (V, E)$, we prove that DeepWalk is actually factorizing a matrix $M \in \mathbb{R}^{|V| \times |V|}$ where each entry M_{ij} is logarithm of the average probability that vertex v_i randomly walks to vertex v_j in fixed steps in last chapter. Figure 3.1a shows the MF-style DeepWalk: factorize matrix M into the product of two low-dimensional matrices $W \in \mathbb{R}^{k \times |V|}$ and $H \in \mathbb{R}^{k \times |V|}$ where $k \ll |V|$. DeepWalk then takes the matrix W as vertex representation.

The matrix-factorization view of DeepWalk inspires us to introduce text information into MF for NE. Figure 3.1b shows the main idea of TADW: factorize matrix M into the product of three matrices: $W \in \mathbb{R}^{k \times |V|}$, $H \in \mathbb{R}^{k \times f_t}$ and text features $T \in \mathbb{R}^{f_t \times |V|}$. Then we concatenate W and HT as $2k$-dimensional representations of vertices.

TADW can learn representations jointly from both network and text information. The representations have less noises and are more consistent. Also, TADW is not task-specific and the representations can be conveniently used for different tasks, such as link prediction, similarity computation and node classification.

We test TADW against several baselines on three datasets. The classification accuracy of TADW outperforms other baselines by at most 2–10% when the ratio of training set ranges from 10–50%. We also test these methods with a semi-supervised classifier, Transductive SVM (TSVM), when training ratio is less than 10%. TADW has a 5–20% advantage than other baselines with 1% training ratio, especially when network information is noisy.

3.2   METHOD: TEXT-ASSOCIATED DEEPWALK

In this section, we first give a brief introduction to low-rank matrix factorization, and then we formulate TADW which learns representations from both network and text information.

3.2.1   LOW-RANK MATRIX FACTORIZATION

Matrix is a common way to represent relational data. An interesting topic for matrix analysis is to figure out the inherent structure of a matrix by a fraction of its entries. One assumption is that matrix $M \in \mathbb{R}^{b \times d}$ admits an approximation of low rank k, where $k \ll \{b, d\}$. Then we can complete the missing entries in matrix M with such a low-rank approximation under this assumption. However, solving a rank constraint optimization is always NP-hard. Therefore, researchers resort to finding matrices $W \in \mathbb{R}^{k \times b}$ and $H \in \mathbb{R}^{k \times d}$ to minimize the loss function $L(M, W^T H)$ with a trace norm constraint, which is further removed by adding a penalty term to the loss function [Yu et al., 2014]. In this chapter, we use square loss function.

Formally, let the observation set of matrix M be Ω. We want to find matrices $W \in \mathbb{R}^{k \times b}$ and $H \in \mathbb{R}^{k \times d}$ to minimize

$$\min_{W,H} \sum_{(i,j)\in\Omega} \left(M_{ij} - (W^T H)_{ij}\right)^2 + \frac{\lambda}{2}\left(||W||_F^2 + ||H||_F^2\right), \tag{3.1}$$

where $|| \cdot ||_F$ means Frobenius norm of the matrix and λ is a harmonic factor to balance two components.

Low-rank matrix factorization completes matrix M only based on the low-rank assumption of M. If items in matrix M have additional features, we can apply inductive matrix completion [Natarajan and Dhillon, 2014] to take advantage of them. Inductive matrix completion utilizes more information of row and column units by incorporating two feature matrices into the objective function. Suppose that we have feature matrices $X \in \mathbb{R}^{f_x \times b}$ and $Y \in \mathbb{R}^{f_y \times d}$ where column i of X and Y are f_x and f_y dimensional feature vectors of unit i, respectively. Our goal is to solve matrices $W \in \mathbb{R}^{k \times f_x}$ and $H \in \mathbb{R}^{k \times f_y}$ to minimize

$$\min_{W,H} \sum_{(i,j)\in\Omega} \left(M_{ij} - (X^T W^T H Y)_{ij}\right)^2 + \frac{\lambda}{2}\left(||W||_F^2 + ||H||_F^2\right). \tag{3.2}$$

Note that, inductive matrix completion is originally proposed to complete gene-disease matrix with gene and disease features [Natarajan and Dhillon, 2014], the goal of which is quite different from that of TADW. Inspired by the idea of inductive matrix completion, we introduce text information into NRL.

3.2.2   TEXT-ASSOCIATED DEEPWALK (TADW)

Given a network $G = (V, E)$ and its corresponding text feature matrix $T \in \mathbb{R}^{f_t \times |V|}$, we propose text-associated DeepWalk (TADW) to learn representation of each vertex $v \in V$ from both network structure G and text features T.

Recall that DeepWalk introduced Skip-Gram [Mikolov et al., 2013b], a widely used distributed word representation method, into the study of social network for the first time to learn vertex representation with network structure. As shown in last chapter, DeepWalk is equivalent to factorize matrix M where $M_{ij} = \log([e_i(A + A^2 + \cdots + A^t)]_j/t)$. Computing an accurate M has the complexity of $O(|V|^3)$ when t gets large. In fact, DeepWalk uses a sampling method based on random walk to avoid explicitly computing accurate matrix M. When DeepWalk samples more walks, the performance will be better while DeepWalk will be less efficient.

In TADW, we find a tradeoff between speed and accuracy: factorize the matrix $M = (A + A^2)/2$. Here, we factorize M instead of $\log M$ for computational efficiency. The reason is that, $\log M$ has much more non-zero entries than M, and the complexity of matrix factorization with square loss is proportional to the number of non-zero elements in matrix M [Yu et al., 2014]. Since most real-world networks are sparse, i.e., $O(E) = O(V)$, computing matrix M takes $O(|V|^2)$ time. If a network is dense, we can even directly factorize matrix A. Our task is to solve matrices $W \in \mathbb{R}^{k \times |V|}$ and $H \in \mathbb{R}^{k \times f_t}$ to minimize

$$\min_{W,H} ||M - W^T H T||_F^2 + \frac{\lambda}{2}(||W||_F^2 + ||H||_F^2). \tag{3.3}$$

For optimizing W and H, we alternately minimize W and H because it is a convex function for either W or H. Though the algorithm may converge to a local minimum instead of the global minimum, TADW works well in practice as shown in the experiments.

Different from low-rank matrix factorization and inductive matrix completion which focus on completing the matrix M, the goal of TADW is to incorporate text features to obtain better network representations. Also, inductive matrix completion obtains matrix M directly from raw data while we artificially build matrix M from the derivation of MF-style DeepWalk. Since both W and HT obtained from TADW can be regarded as low-dimensional representations of vertices, we build a unified $2k$-dimensional matrix for network representations by concatenating them. In experiments, we will show that the unified representation significantly outperforms a naive combination of network representations and text features (i.e., the matrix T).

3.2.3   COMPLEXITY ANALYSIS

In TADW, the procedure of computing M takes $O(|V|^2)$ time. We use the fast procedure introduced by Yu et al. [2014] to solve the optimization problem in Eq. (3.3). The complexity of each iteration of minimizing W and H is $O(\text{nnz}(M)k + |V|f_t k + |V|k^2)$ where nnz(·) indicates the number of non-zero entries. For comparison, the complexity of traditional matrix factorization, i.e., optimization problem in Eq. (3.1), is $O(\text{nnz}(M)k + |V|k^2)$. In fact, the optimization usually converges within 10 iterations in the experiments.

3.3   EMPIRICAL ANALYSIS

We use multi-class vertex classification to evaluate the quality of NE. Formally, we regard low-dimensional representation $\mathcal{R} = \{r_1, r_2, \ldots, r_{|V|}\}$ as node features. Our task is to predict the labels of unlabeled set U with labeled set L based on vertex features $\mathcal{R}$.

A number of classifiers in machine learning can deal with this task. We select SVM and TSVM for supervised and semi-supervised learning and testing, respectively. Note that, since the representation learning procedure ignores vertex labels in training set, representation learning is unsupervised.

We evaluate TADW with five baseline methods of representation learning using three publicly available datasets. We learn representations from the links or citations between the documents as well as the term frequency-inverse document frequency (TFIDF) matrices of these documents.

3.3.1   DATASETS

Cora [McCallum et al., 2000] contains 2,708 machine learning papers from 7 classes and 5, 429 links between them. The links are citation relationships between the documents. Each document is described by a binary vector of 1,433 dimensions indicating the presence of the corresponding word.

Citeseer [McCallum et al., 2000] contains 3,312 publications from six classes and 4,732 links between them. Similar to Cora, the links are citation relationships between the documents and each paper is described by a binary vector of 3,703 dimensions.

Wiki [Sen et al., 2008] contains 2,405 documents from 19 classes and 17,981 links between them. The TFIDF matrix of this dataset has 4,973 columns.

The documents in Cora and Citeseer are short texts generated from titles and abstracts. Stop words and all words with document frequency less than 10 are removed. Each document has 18 or 32 words on average correspondingly. The documents in Wiki are long texts. We remove all documents which have no connection in the network. Each document has 640 words on average. We regard the networks as undirected graphs.

3.3.2   TADW SETTINGS

For all three datasets, we reduce the dimension of word vectors to 200 via SVD decomposition of the TFIDF matrix, and obtain text feature matrices $T \in \mathbb{R}^{200 \times |V|}$. The preprocessing will reduce the number of parameters in matrix H. We also take text feature matrix T as a content-only baseline. We select $k = 80$ and $\lambda = 0.2$ for Cora and Citeseer datasets; $k = 100, 200$, and $\lambda = 0.2$ for Wiki dataset. Note that the dimension of representation vectors from TADW is $2k$.

3.3.3   BASELINE METHODS

DeepWalk. DeepWalk [Perozzi et al., 2014] is a network-only representation learning method. We set parameters as follows: walks per vertex $\gamma = 80$ and window size $t = 10$ which are the same with those in the original paper. We choose representation dimension $k = 100$ for Cora and Citeseer and $k = 200$ for Wiki which are the lengths with best performance between 50 and 200.

We also evaluate the performance of MF-style DeepWalk by solving Eq. (3.1) and concatenate W and H as vertex representations. The result is competitive with DeepWalk. Hence we only report the performance of original DeepWalk.

PLSA. We use PLSA [Hofmann, 1999] to train a topic model from the TFIDF matrix by regarding each vertex as a document. Hence, PLSA is a content-only baseline. PLSA estimates topic distribution of documents and word distribution of topics via EM algorithm. We use topic distribution of documents as vertex representations.

Text Features. We use text feature matrix $T \in \mathbb{R}^{200 \times |V|}$ as a 200-dimensional representation. The method of Text Features is a content-only baseline.

Naive Combination. We can simply concatenate the vectors from both Text Features and DeepWalk for network representations. It has a length of 300 for Cora and Citeseer and 400 for Wiki.

NetPLSA. [Mei et al., 2008] proposed learning topic distributions of documents by considering links between documents as a network regularization that linked documents should share similar topic distributions. We use the network-enhanced topic distribution of documents as network representations. NetPLSA can be regarded as an NE method considering both network and text information. We set topic numbers to 160 for Cora and Citeseer, and 200 for Wiki.

3.3.4   CLASSIFIERS AND EXPERIMENT SETUP

For a supervised classifier, we use linear SVM implemented by Liblinear [Fan et al., 2008]. For semi-supervised classifiers, we use transductive SVM implemented by SVM-Light [Joachims, 1999]. We use linear kernel for TSVM. We train a one-vs-rest classifier for each class and select the classes with maximum scores in linear SVM and transductive SVM.

We take representations of vertices as features to train classifiers, and evaluate classification accuracy with different training ratios. The training ratio varies from 10–50% for linear SVM and 1–10% for TSVM. For each training ratio, we randomly select documents as training set and the remaining documents as test set. We repeat the trial for 10 times and report the average accuracy.

Table 3.1: Evaluation results on Cora dataset

Classifier	Transductive SVM				SVM				
% Labeled nodes	1%	3%	7%	10%	10%	20%	30%	40%	50%
DeepWalk	62.9	68.3	72.2	72.8	76.4	78.0	79.5	80.5	81.0
PLSA	47.7	51.9	55.2	60.7	57.0	63.1	65.1	66.6	67.6
Text features	33.0	43.0	57.1	62.8	58.3	67.4	71.1	73.3	74.0
Naive combination	67.4	70.6	75.1	77.4	76.5	80.4	82.3	83.3	84.1
NetPLSA	65.7	67.9	74.5	77.3	80.2	83.0	84.0	84.9	85.4
TADW	**72.1**	**77.0**	**79.1**	**81.3**	**82.4**	**85.0**	**85.6**	**86.0**	**86.7**

Table 3.2: Evaluation results on Citeseer dataset

Classifier	Transductive SVM				SVM				
% Labeled nodes	1%	3%	7%	10%	10%	20%	30%	40%	50%
DeepWalk	—	—	49.0	52.1	52.4	54.7	56.0	56.5	57.3
PLSA	45.2	49.2	53.1	54.6	54.1	58.3	60.9	62.1	62.6
Text features	36.1	49.8	57.7	62.1	58.3	66.4	69.2	71.2	72.2
Naive combination	39.0	45.7	58.9	61.0	61.0	66.7	69.1	70.8	72.0
NetPLSA	45.4	49.8	52.9	54.9	58.7	61.6	63.3	64.0	64.7
TADW	**63.6**	**68.4**	**69.1**	**71.1**	**70.6**	**71.9**	**73.3**	**73.7**	**74.2**

3.3.5 EXPERIMENTAL RESULTS AND ANALYSIS

Tables 3.1, 3.2, and 3.3 show classification accuracies on Cora, Citeseer, and Wiki datasets. Here "-" indicates TSVM cannot converge in 12 hours because of low quality of representation (TSVM can always converge in 5 minutes for TADW). We did not show the results of semi-supervised learning on Wiki dataset because supervised SVM has already attained a competitive and even better performance with small training ratio on this dataset. Thus, we only report the results of supervised SVM for Wiki. Wiki has much more classes than the other two datasets, which requires more data for sufficient training, hence we set the minimum training ratio to 3%. From these tables, we have the following observations:

(1) TADW consistently outperforms all the other baselines on all three datasets. Furthermore, TADW can beat other baselines with 50% less training data on Cora and Citeseer datasets. These experiments demonstrate that TADW is effective and robust.

(2) TADW has more significant improvement for semi-supervised learning. TADW outperforms the best baseline, i.e., naive combination, by 4% on Cora and 10–20% on Citeseer.

Table 3.3: Evaluation results on Wiki dataset

Classifier	SVM						
% Labeled nodes	3%	7%	10%	20%	30%	40%	50%
DeepWalk	48.4	56.6	59.3	64.3	66.2	68.1	68.8
PLSA	58.3	66.5	69.0	72.5	74.7	75.5	76.0
Text features	46.7	60.8	65.1	72.9	75.6	77.1	77.4
Naive combination	48.7	62.6	66.3	73.0	75.2	77.1	78.6
NetPLSA	56.3	64.6	67.2	70.6	71.7	71.9	72.3
TADW ($k = 100$)	59.8	68.2	71.6	75.4	77.3	77.7	79.2
TADW ($k = 200$)	60.4	69.9	72.6	77.3	79.2	79.9	80.3

This is because the quality of network representations is poor on Citeseer, while TADW is more robust for learning from noisy data than naive combination.

(3) TADW has an encouraging performance when training ratio is small. The accuracies of most baselines drop quickly as training ratio decreases because their vertex representations are much noisy and inconsistent for training and testing. Instead, since TADW learns representation jointly from both network and text information, the representations have less noises and are more consistent.

These observations demonstrate the high quality of representations generated by TADW. Moreover, TADW is not task-specific and the representations can be conveniently used for different tasks, such as link prediction, similarity computation, and vertex classification.

3.3.6   CASE STUDY

To better understand the effectiveness of node attribute information for NRL, we present an example in Cora dataset. As shown in Table 3.4, using representations generated by DeepWalk and TADW, we find five most similar documents ranked by cosine similarity.

We find that, all these documents are cited by the target document. However, three of the five documents found by DeepWalk have different class labels while the first four documents found by TADW have the same label "Theory." This indicates the necessity of incorporating node features into NRL.

The 5th document found by DeepWalk also shows another limitation of considering only network information. "MLC Tutorial A Machine Learning library of C classes" (MLC for short) is a document describing a general toolbox, which may be cited by many works in different topics. Once some of these works cite the target document as well, DeepWalk will tend to give the target document a similar representation with MLC even though they are totally on different topics.

Table 3.4: Five most similar documents ranked by DeepWalk and TADW

Target document	
Title	**Class label**
Irrelevant features and the subset selection problem	Theory
Top 5 nearest documents by DeepWalk	
Title	**Class label**
Feature selection methods for classifications	Neural network
Automated model selection	Rule learning
Compression-based feature subset selection	Theory
Induction of condensed determinations	Case based
MLC Tutorial A machine learning library of C classes	Theory
Top 5 nearest documents by TADW	
Title	**Class label**
Feature subset selection as search with probabilistic estimates	Theory
Compression-based feature subset selection	Theory
Selection of relevant features in machine learning	Theory
NP-completeness of searches for smallest possible feature sets	Theory
Feature subset selection using a genetic algorithm	Genetic algorithms

3.4   FURTHER READING

NE algorithms [Chen et al., 2007, Perozzi et al., 2014, Tang and Liu, 2009, 2011] introduced in the previous chapter are hard to generalize to deal with other features of nodes trivially. To the best of our knowledge, little work has been devoted to consider node attributes in NRL before TADW. There are some topic models, such as NetPLSA [Mei et al., 2008], considering both networks and text information for topic modeling, in which we can represent each node with a topic distribution. However, the representation capacity of NetPLSA is relatively low and thus yields suboptimal performance on downstream tasks such as node classification.

There have been a number of follow-up works of TADW. For example, HSCA (Homophily, Structure, and Content Augmented network representation learning) [Zhang et al., 2016] was proposed to emphasize homophily between connected nodes by introducing additional regularization terms to TADW. Specifically, we denote the output embedding of TADW as $R = [W \parallel HT]$ where each row of R is the embedding of a node. Then the regularization term can be written as

$$Reg(W, H) = \sum_{(v_i, v_j) \in E} ||r_i - r_j||^2 = \sum_{(v_i, v_j) \in E} (||w_i - w_j||^2 + ||HT_i - HT_j||^2), \qquad (3.4)$$

where r_i, w_i, and HT_i are the i-th row of matrix R, W, and HT. The regularization term will be jointly optimized with the loss function in Eq. (3.3). To this end, HSCA can better preserve the homophily property of a network.

Sun et al. [2016] regard text content as a special kind of nodes, and propose context-enhanced NE (CENE) through leveraging both structural and textural information to learn NEs. Inspired by the merits of SDNE [Wang et al., 2016a], DANE [Gao and Huang, 2018] employs deep neural architectures for attributed NE. Other extensions include the modeling of additional label information [Pan et al., 2016, Wang et al., 2016b], dynamic environments [Li et al., 2017b], attribute proximity [Liao et al., 2018], attributed signed networks [Wang et al., 2017e], etc. Readers are encouraged to take a look at these papers for a deeper insight into the problem.

Part of this chapter was published in our IJCAI15 conference paper by Yang et al. [2015].

Revisiting Attributed Network Embedding: A GCN-Based Perspective

Attributed graph embedding, which learns vector representations from graph topology and node features, is a challenging task for graph analysis. Though TADW and other methods mentioned in Chapter 3 give generally good performances, methods based on graph convolutional networks (GCNs) have shown their superiority on this task recently. In this chapter, we will revisit attributed NE from a GCN-based perspective. We will start by a brief introduction to GCN and GCN-based NE methods. Afterward, we will present three major drawbacks in existing GCN-based methods and introduce Adaptive Graph Encoder (AGE), a novel attributed graph embedding framework, to address these issues. Finally, we conduct experiments using four public benchmark datasets to demonstrate the effectiveness of AGE. Experimental results on node clustering and link prediction tasks show that AGE consistently outperforms state-of-the-art graph embedding methods.

4.1   GCN-BASED NETWORK EMBEDDING

There has been a surge of approaches that focus on deep learning on graphs. Specifically, approaches from the family of GCNs [Kipf and Welling, 2017] have made great progress in many graph learning tasks [Zhou et al., 2018] and strengthen the representation power of graph embedding algorithms. We will introduce GCN and other GCN-based network embedding methods in this section.

4.1.1   GRAPH CONVOLUTIONAL NETWORK (GCN)

The goal of GCN is to learn a function of features on a graph, with network structure $G = (V, E)$ and node feature $x_i \in \mathbb{R}^{d_f}$ for each node $v_i \in V$ as inputs. GCN is composed of multiple layers of graph convolution operations which update node embeddings by aggregating their neighborhood information.

Formally, denote the adjacency matrix as A and degree matrix as D, the node embeddings $Z^{(t)}$ after t layers can be computed recursively as

$$Z^{(t)} = f(Z^{(t-1)}, A) = \sigma(AZ^{(t-1)}W^{(t)}), \qquad (4.1)$$

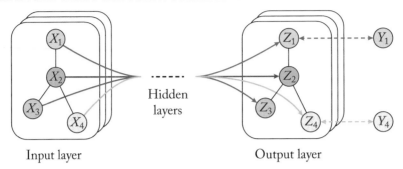

Figure 4.1: The architecture of graph convolutional network (GCN) [Kipf and Welling, 2017].

where $\sigma(\cdot)$ is a nonlinear activation function (*e.g., ReLU or Sigmoid*), $Z^{(0)} \in \mathbb{R}^{|V| \times d_f}$ is the initial node features and $W^{(t)}$ is the weight matrix to be learned in t-th layer.

Note that stacking the operation in Eq. (4.1) could lead to numerical instabilities and exploding/vanishing gradients. Therefore, GCN employed renormalization trick and modified Eq. (4.1) as

$$Z^{(t)} = f(Z^{(t-1)}, A) = \sigma(\tilde{D}^{-\frac{1}{2}} \tilde{A} \tilde{D}^{-\frac{1}{2}} Z^{(t-1)} W^{(t)}), \qquad (4.2)$$

where $\tilde{A} = A + I$ is the adjacency matrix with self loops and $\tilde{D}$ is the corresponding degree matrix of $\tilde{A}$.

After T layers of graph convolution operations, we use the embedding matrix at last layer $Z^{(T)}$ and a readout function to compute the final output matrix Z:

$$Z = \text{Readout}(Z^{(T)}), \qquad (4.3)$$

where the readout function can be any neural models, such as MLP.

Finally, as a semi-supervised algorithm, GCN forced the dimension of output Z to be equal with the number of labels and then employed a softmax function to normalize the output for label prediction. The loss function can be written as

$$\mathcal{L} = - \sum_{l \in y_L} \sum_k Y_{lk} \ln Z_{lk}, \qquad (4.4)$$

where y_L is the set of node indices that have observed labels. Figure 4.1 shows the architecture of GCN.

4.1.2   ATTRIBUTED NETWORK EMBEDDING BASED ON GCN

GCN was originally proposed for semi-supervised task-specific graph modeling. Several follow-up work adopted GCN for unsupervised attributed NE without label information. These GCN-based methods can be categorized into two groups by their optimization objectives.

Reconstruct the adjacency matrix. This kind of approaches forces the learned embeddings to recover their localized neighborhood structure. Graph autoencoder (GAE) and variational graph autoencoder (VGAE) [Kipf and Welling, 2016] learn node embeddings by using GCN as the encoder, then decode by inner product with cross-entropy loss. Formally, GAE aims to minimize the distance between adjacency matrix A and the reconstructed one $\sigma(ZZ^T)$. As variants of GAE (VGAE), Pan et al. [2018] exploits adversarially regularized method to learn more robust node embeddings. Wang et al. [2019a] further employs graph attention networks [Veličković et al., 2018] to differentiate the importance of the neighboring nodes to a target node.

Reconstruct the feature matrix. This kind of models is autoencoders for the node feature matrix while the adjacency matrix merely serves as a filter. Wang et al. [2017a] leverages marginalized denoising autoencoder to disturb the structure information. To build a symmetric graph autoencoder, Park et al. [2019] proposes Laplacian sharpening as the counterpart of Laplacian smoothing in the encoder. The authors claim that Laplacian sharpening is a process that makes the reconstructed feature of each node away from the centroid of its neighbors to avoid over-smoothing. However, as we will show in the next section, there exists high-frequency noises in raw node features, which harm the quality of learned embeddings.

4.1.3   DISCUSSIONS

Most of existing GCN-based NE methods are based on graph autoencoder (GAE) and variational graph autoencoder (VGAE) [Kipf and Welling, 2016]. As shown in Fig. 4.2, they comprise a GCN encoder and a reconstruction decoder. Nevertheless, these GCN-based methods have three major drawbacks:

First, a GCN encoder consists of multiple graph convolutional layers, and each layer contains a graph convolutional filter (H in Fig. 4.2), a weight matrix (W_1, W_2 in Fig. 4.2) and an activation function. However, previous work [Wu et al., 2019a] demonstrates that the entanglement of the filters and weight matrices provides no performance gain for semi-supervised graph representation learning, and even harms training efficiency since it deepens the paths of back-propagation. In this chapter, we further extend this conclusion to unsupervised scenarios by controlled experiments, showing that our disentangled architecture performs better and more robust than entangled models (Section 4.3.6).

Second, considering the graph convolutional filters, previous research [Li et al., 2018] shows in theory that they are actually Laplacian smoothing filters [Taubin, 1995] applied on the feature matrix for low-pass denoising. But we show that existing graph convolutional filters are not optimal low-pass filters since they cannot filter out noises in some high-frequency intervals. Thus, they cannot reach the best smoothing effect (Section 4.3.8).

Third, we also argue that training objectives of these algorithms (either reconstructing the adjacency matrix [Pan et al., 2018, Wang et al., 2019a] or feature matrix [Park et al., 2019, Wang et al., 2017a]) are not compatible with real-world applications. To be specific, reconstructing

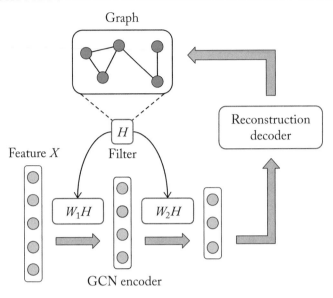

Figure 4.2: The architecture of graph autoencoder [Kipf and Welling, 2016]. The components we argue about are marked in red blocks: entanglement of the filters and weight matrices, design of the filters, and the reconstruction loss.

adjacency matrix literally sets the adjacency matrix as the ground truth pairwise similarity, while it is not proper for the lack of feature information. Recovering the feature matrix, however, will force the model to remember high-frequency noises in features, and thus be inappropriate as well.

Motivated by such observations, we propose Adaptive Graph Encoder (AGE), a unified framework for attributed graph embedding in this chapter. To disentangle the filters and weight matrices, AGE consists of two modules: (1) A well-designed nonparametric Laplacian smoothing filter to perform low-pass filtering in order to get smoothed features. (2) An adaptive encoder to learn more representative node embeddings. To replace the reconstruction training objectives, we employ adaptive learning [Chang et al., 2017] in this step, which selects training samples from the pairwise similarity matrix and finetunes the embeddings iteratively.

4.2   METHOD: ADAPTIVE GRAPH ENCODER

In this section, we first formalize the embedding task on attributed graphs. Then we present our proposed AGE algorithm. Specifically, we first design an effective graph filter to perform Laplacian smoothing on node features. Given the smoothed node features, we further develop a simple node representation learning module based on adaptive learning [Chang et al., 2017].

Finally, the learned node embeddings can be used for downstream tasks such as node clustering and link prediction.

4.2.1 PROBLEM FORMALIZATION

Given an attributed graph $G = (V, E, X)$, where $V = \{v_1, v_2, \cdots, v_n\}$ is the vertex set with n nodes in total, E is the edge set, and $X = [x_1, x_2, \cdots, x_n]^T$ is the feature matrix. The topology structure of graph G can be denoted by an adjacency matrix $A = \{a_{ij}\} \in \mathbb{R}^{n \times n}$, where $a_{ij} = 1$ if $(v_i, v_j) \in E$, indicating there is an edge from node v_i to node v_j. $D = \text{diag}(d_1, d_2, \cdots, d_n) \in \mathbb{R}^{n \times n}$ denotes the degree matrix of A, where $d_i = \sum_{v_j \in V} a_{ij}$ is the degree of node v_i. The graph Laplacian matrix is defined as $L = D - A$.

The purpose of attributed graph embedding is to map nodes to low-dimensional embeddings. We take Z as the embedding matrix and the embeddings should preserve both the topological structure and feature information of graph G.

4.2.2 OVERALL FRAMEWORK

The framework of our model is shown in Fig. 4.3. It consists of two parts: a Laplacian smoothing filter and an adaptive encoder.

- **Laplacian Smoothing Filter**: The designed filter H serves as a low-pass filter to denoise the high-frequency components of the feature matrix X. The smoothed feature matrix $\tilde{X}$ is taken as input of the adaptive encoder.

- **Adaptive Encoder**: To get more representative node embeddings, this module builds a training set by adaptively selecting node pairs which are highly similar or dissimilar. Then the encoder is trained in a supervised manner.

After the training process, the learned node embedding matrix Z is used for downstream tasks.

4.2.3 LAPLACIAN SMOOTHING FILTER

The basic assumption for graph learning is that nearby nodes on the graph should be similar, thus node features are supposed to be *smooth* on the graph manifold. In this section, we first explain what *smooth* means. Then we give the definition of the generalized Laplacian smoothing filter and show that it is a smoothing operator. Finally, we will show how to design an optimal Laplacian smoothing filter.

Analysis of Smooth Signals. We start with interpreting *smooth* from the perspective of graph signal processing. Take $x \in \mathbb{R}^n$ as a graph signal where each node is assigned with a scalar. Denote the filter matrix as H. To measure the smoothness of graph signal x, we can calculate the *Rayleigh quotient* [Horn and Johnson, 2012] over the graph Laplacian L and x:

$$R(L, x) = \frac{x^T L x}{x^T x} = \frac{\sum_{(i,j) \in E} (x_i - x_j)^2}{\sum_{i \in V} x_i^2}. \tag{4.5}$$

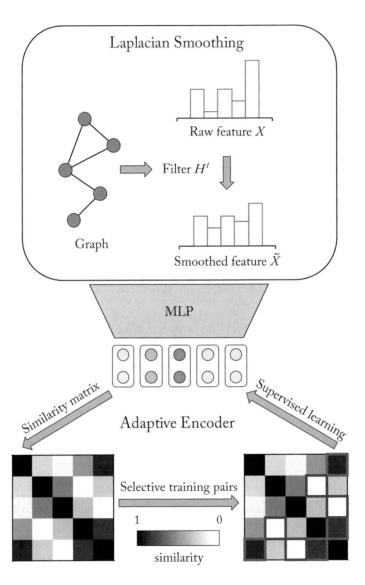

Figure 4.3: Our AGE framework. Given the raw feature matrix X, we first perform t-layer Laplacian smoothing using filter H^t to get the smoothed feature matrix $\tilde{X}$ (**Top**). Then the node embeddings are encoded by the adaptive encoder which utilizes the adaptive learning strategy: (1) Calculate the pairwise node similarity matrix. (2) Select positive and negative training samples of high confidence (red and green squares). (3) Train the encoder by a supervised loss (**Bottom**).

This quotient is actually the normalized variance score of x. As stated above, *smooth* signals should assign similar values on neighboring nodes. Consequently, signals with lower *Rayleigh quotient* are assumed to be smoother.

Consider the eigendecomposition of graph Laplacian $L = U\Lambda U^{-1}$, where $U \in \mathbb{R}^{n \times n}$ comprises eigenvectors and $\Lambda = \mathrm{diag}(\lambda_1, \lambda_2, \cdots, \lambda_n)$ is a diagonal matrix of eigenvalues. Then the smoothness of eigenvector u_i is given by

$$R(L, u_i) = \frac{u_i^T L u_i}{u_i^T u_i} = \lambda_i. \tag{4.6}$$

Equation (4.6) indicates that smoother eigenvectors are associated with smaller eigenvalues, which means lower frequencies. Thus, we decompose signal x on the basis of L based on Eqs. (4.5) and (4.6):

$$x = Up = \sum_{i=1}^{n} p_i u_i, \tag{4.7}$$

where p_i is the coefficient of eigenvector u_i. Then the smoothness of x is actually

$$R(L, x) = \frac{x^T L x}{x^T x} = \frac{\sum_{i=1}^{n} p_i^2 \lambda_i}{\sum_{i=1}^{n} p_i^2}. \tag{4.8}$$

Therefore, to get smoother signals, the goal of our filter is filtering out high-frequency components while preserving low-frequency components. Because of its high computational efficiency and convincing performance, Laplacian smoothing filters [Taubin, 1995] are often utilized for this purpose.

Generalized Laplacian Smoothing Filter. As stated by Taubin [1995], the generalized Laplacian smoothing filter is defined as

$$H = I - kL, \tag{4.9}$$

where k is real-valued. Employ H as the filter matrix, the filtered signal $\tilde{x}$ is present by

$$\tilde{x} = Hx = U(I - k\Lambda)U^{-1}Up = \sum_{i=1}^{n}(1 - k\lambda_i)p_i u_i = \sum_{i=1}^{n} p_i' u_i. \tag{4.10}$$

Hence, to achieve low-pass filtering, the frequency response function $1 - k\lambda$ should be a decrement and non-negative function. Stacking up t Laplacian smoothing filters, we denote the filtered feature matrix $\tilde{X}$ as

$$\tilde{X} = H^t X. \tag{4.11}$$

Note that the filter is nonparametric at all.

The Choice of k**.** In practice, with the renormalization trick $\tilde{A} = I + A$, we employ the symmetric normalized graph Laplacian

$$\tilde{L}_{sym} = \tilde{D}^{-\frac{1}{2}} \tilde{L} \tilde{D}^{-\frac{1}{2}}, \tag{4.12}$$

where $\tilde{D}$ and $\tilde{L}$ are degree matrix and Laplacian matrix corresponding to $\tilde{A}$. Then the filter becomes

$$H = I - k\tilde{L}_{sym}. \tag{4.13}$$

Notice that if we set $k = 1$, the filter becomes the GCN filter.

For selecting optimal k, the distribution of eigenvalues $\tilde{\Lambda}$ (obtained from the decomposition of $\tilde{L}_{sym} = \tilde{U}\tilde{\Lambda}\tilde{U}^{-1}$) should be carefully discovered.

The smoothness of $\tilde{x}$ is

$$R(L, \tilde{x}) = \frac{\tilde{x}^T L \tilde{x}}{\tilde{x}^T \tilde{x}} = \frac{\sum_{i=1}^{n} p_i'^2 \lambda_i}{\sum_{i=1}^{n} p_i'^2}. \tag{4.14}$$

Thus, $p_i'^2$ should decrease as λ_i increases. We denote the maximum eigenvalue as $\lambda_{\max}$. Theoretically, if $k > 1/\lambda_{\max}$, the filter is not low-pass in the $(1/k, \lambda_{\max}]$ interval because $p_i'^2$ increases in this interval; otherwise, if $k < 1/\lambda_{\max}$, the filter cannot denoise all the high-frequency components. Consequently, $k = 1/\lambda_{\max}$ is the optimal choice.

It has been proved that the range of Laplacian eigenvalues is between 0 and 2 [Chung and Graham, 1997], hence GCN filter is not low-pass in the $(1, 2]$ interval. Some work [Wang et al., 2019a] accordingly chooses $k = 1/2$. However, our experiments show that after renormalization, the maximum eigenvalue λ_{max} will shrink to around $3/2$, which makes $1/2$ not optimal as well. In experiments, we calculate $\lambda_{\max}$ for each dataset and set $k = 1/\lambda_{\max}$. We further analyze the effects of different k values (Section 4.3.8).

4.2.4   ADAPTIVE ENCODER

Filtered by t-layer Laplacian smoothing, the output features are smoother and preserve abundant attribute information.

To learn better node embeddings from the smoothed features, we need to find an appropriate unsupervised optimization objective. To this end, we manage to utilize pairwise node similarity inspired by Deep Adaptive Learning [Chang et al., 2017]. For attributed graph embedding task, the relationship between two nodes is crucial, which requires the training targets to be suitable similarity measurements. GAE-based methods usually choose the adjacency matrix as true labels of node pairs. However, we argue that the adjacency matrix only records one-hop structure information, which is insufficient. Meanwhile, we address that the similarity of smoothed features or trained embeddings are more accurate since they incorporate structure and features together. To this end, we adaptively select node pairs of high similarity as positive training samples, while those of low similarity as negative samples.

Given filtered node features $\tilde{X}$, the node embeddings are encoded by linear encoder f:

$$Z = f(\tilde{X}; W) = \tilde{X}W, \tag{4.15}$$

where W is the weight matrix. We then scale the embeddings to the $[0, 1]$ interval by min-max scaler for variance reduction. To measure the pairwise similarity of nodes, we utilize cosine function to implement our similarity metric. The similarity matrix S is given by

$$S = \frac{ZZ^T}{\|Z\|_2^2}. \tag{4.16}$$

Next, we describe our training sample selection strategy in detail.

Training Sample Selection. After calculating the similarity matrix, we rank the pairwise similarity sequence in the descending order. Here r_{ij} is the rank of node pair (v_i, v_j). Then we set the maximum rank of positive samples as r_{pos} and the minimum rank of negative samples as r_{neg}. Therefore, the generated label of node pair (v_i, v_j) is

$$l_{ij} = \begin{cases} 1 & r_{ij} \leq r_{pos} \\ 0 & r_{ij} > r_{neg} \\ \text{None} & \text{otherwise} \end{cases}. \tag{4.17}$$

In this way, a training set with r_{pos} positive samples and $n^2 - r_{neg}$ negative samples is constructed. Specially, for the first time we construct the training set, since the encoder is not trained, we directly employ the smoothed features for initializing S:

$$S = \frac{\tilde{X}\tilde{X}^T}{\|\tilde{X}\|_2^2}. \tag{4.18}$$

After construction of the training set, we can train the encoder in a supervised manner. In real-world graphs, there are always far more dissimilar node pairs than positive pairs, so we select more than r_{pos} negative samples in the training set. To balance positive/negative samples, we randomly choose r_{pos} negative samples in every epoch. The balanced training set is denoted by $\mathcal{O}$. Accordingly, our cross entropy loss is given by

$$\mathcal{L} = \sum_{(v_i, v_j) \in \mathcal{O}} -l_{ij} \log(s_{ij}) - (1 - l_{ij}) \log(1 - s_{ij}). \tag{4.19}$$

Thresholds Update. Inspired by the idea of curriculum learning [Bengio et al., 2009], we design a specific update strategy for r_{pos} and r_{neg} to control the size of training set. At the beginning of training process, more samples are selected for the encoder to find rough cluster patterns. After that, samples with higher confidence are remained for training, forcing the encoder to capture refined patterns. In practice, r_{pos} decreases while r_{neg} increases linearly as the training procedure goes on. We set the initial threshold as r_{pos}^{st} and r_{neg}^{st}, together with the final threshold as r_{pos}^{ed} and r_{neg}^{ed}. We have $r_{pos}^{ed} \leq r_{pos}^{st}$ and $r_{neg}^{ed} \geq r_{neg}^{st}$. Suppose the thresholds are updated T times; we present the update strategy as

$$r'_{pos} = r_{pos} + \frac{r_{pos}^{ed} - r_{pos}^{st}}{T}, \tag{4.20}$$

$$r'_{neg} = r_{neg} + \frac{r_{neg}^{ed} - r_{neg}^{st}}{T}. \tag{4.21}$$

As the training process goes on, every time the thresholds are updated, we reconstruct the training set and save the embeddings. For node clustering, we perform Spectral Clustering [Ng et al., 2002] on the similarity matrices of saved embeddings, and select the best epoch by Davies–Bouldin index (DBI) [Davies and Bouldin, 1979], which measures the clustering quality without label information. For link prediction, we select the best performed epoch on validation set. Algorithm 0 presents the overall procedure of computing the embedding matrix Z.

Algorithm 4.2 Adaptive Graph Encoder

Require: Adjacency matrix A, feature matrix X, filter layer number t, iteration number max_$iter$, and threshold update times T
Ensure: Node embedding matrix Z
  1: Obtain graph Laplacian $\tilde{L}_{sym}$ from Eq. (4.12);
  2: $k \leftarrow 1/\lambda_{\max}$;
  3: Get filter matrix H from Eq. (4.13);
  4: Get smoothed feature matrix $\tilde{X}$ from Eq. (4.11);
  5: Initialize similarity matrix S and training set $\mathcal{O}$ by Eq. (4.18);
  6: **for** $iter = 1$ to max_$iter$ **do**
  7:     Compute Z with Eq. (4.15);
  8:     Train the adaptive encoder with loss in Eq. (4.19);
  9:     **if** $iter$ mod (max_$iter/T$) $== 0$ **then**
 10:         Update thresholds with Eq. (4.20) and (4.21);
 11:         Calculate the similarity matrix S with Eq. (4.16);
 12:         Select training samples from S by Eq. (4.17);
 13:     **end if**
 14: **end for**

Table 4.1: Dataset statistics

Dataset	# Nodes	# Edges	# Features	# Classes
Cora	2,708	5,429	1,433	7
Citeseer	3,327	4,732	3,703	6
Wiki	2,405	17,981	4,973	17
Pubmed	19,717	44,338	500	3

4.3   EMPIRICAL ANALYSIS

We evaluate the benefits of AGE against a number of state-of-the-art graph embedding approaches on node clustering and link prediction tasks. In this section, we introduce the benchmark datasets, baseline methods, evaluation metrics, and parameter settings. Then we show and analyze the results of our experiments. Besides the main experiments, we also conduct auxiliary experiments to answer the following hypotheses.

H1: Entanglement of the filters and weight matrices has no improvement for embedding quality.

H2: Our adaptive learning strategy is effective compared to reconstruction losses, and each mechanism has its own contribution.

H3: $k = 1/\lambda_{\max}$ is the optimal choice for Laplacian smoothing filters.

4.3.1   DATASETS

We conduct node clustering and link prediction experiments on four widely used network datasets (Cora, Citeseer, Pubmed [Sen et al., 2008], and Wiki [Yang et al., 2015]). Features in Cora and Citeseer are binary word vectors, while in Wiki and Pubmed, nodes are associated with tf-idf weighted word vectors. The statistics of the four datasets are shown in Table 4.1.

4.3.2   BASELINE METHODS

For attributed graph embedding methods, we include five baseline algorithms in our comparisons:

GAE and VGAE [Kipf and Welling, 2016] combine graph convolutional networks with the (variational) autoencoder for representation learning.

ARGA and ARVGA [Pan et al., 2018] add adversarial constraints to GAE and VGAE, respectively, enforcing the latent representations to match a prior distribution for robust node embeddings.

GALA [Park et al., 2019] proposes a symmetric graph convolutional autoencoder recovering the feature matrix. The encoder is based on Laplacian smoothing while the decoder is based on Laplacian sharpening.

On the node clustering task, we compare our model with eight more algorithms. The baselines can be categorized into three groups.

(1) Methods using features only. Kmeans [Lloyd, 1982] and Spectral Clustering [Ng et al., 2002] are two traditional clustering algorithms. Spectral-F takes the cosine similarity of node features as input.

(2) Methods using graph structure only. Spectral-G is Spectral Clustering with the adjacency matrix as the input similarity matrix. DeepWalk [Perozzi et al., 2014] learns node embeddings by using Skip-Gram on generated random walk paths on graphs.

(3) Methods using both features and graph. TADW [Yang et al., 2015] interprets DeepWalk as matrix factorization and incorporates node features under the DeepWalk framework. MGAE [Wang et al., 2017a] is a denoising marginalized graph autoencoder. Its training objective is reconstructing the feature matrix. AGC [Zhang et al., 2019c] exploits high-order graph convolution to filter node features. The number of graph convolution layers are selected for different datasets. DAEGC [Wang et al., 2019a] employs graph attention network to capture the importance of the neighboring nodes, then co-optimize reconstruction loss and KL-divergence-based clustering loss.

For representation learning algorithms including DeepWalk, TADW, GAE, and VGAE which do not specify on the node clustering problem, we apply Spectral Clustering on their learned representations. For other works that conduct experiments on benchmark datasets, the original results in the papers are reported.

AGE variants. We consider four variants of AGE to compare various optimization objectives. The Laplacian smoothing filters in these variants are the same, while the encoder of LS+RA aims at reconstructing the adjacency matrix. LS+RX, respectively, reconstructs the feature matrix. LS only preserves the Laplacian smoothing filter, the smoothed features are taken as node embeddings. AGE is the proposed model with adaptive learning.

4.3.3   EVALUATION METRICS AND PARAMETER SETTINGS

To measure the performance of node clustering methods, we employ two metrics: Normalized Mutual Information (NMI) and Adjusted Rand Index (ARI) [Gan et al., 2007]. For link prediction, we partition the datasets following GAE, and report Area Under Curve (AUC) and Average Precision (AP) scores. For all the metrics, a higher value indicates better performance.

For the Laplacian smoothing filter, we find the maximum eigenvalues of the four datasets are all around $3/2$. Thus, we set $k = 2/3$ universally. For the adaptive encoder, we train the MLP encoder for 400 epochs with a 0.001 learning rate by the Adam optimizer [Kingma and

Table 4.2: Hyperparameter settings, where n is number of nodes in the dataset

Dataset	t	r^{st}_{pos}/n^2	r^{ed}_{pos}/n^2	r^{st}_{neg}/n^2	r^{ed}_{neg}/n^2
Cora	8	0.0110	0.0010	0.1	0.5
Citeseer	3	0.0015	0.0010	0.1	0.5
Wiki	1	0.0011	0.0010	0.1	0.5
Pubmed	35	0.0013	0.0010	0.7	0.8

Ba, 2015]. The encoder consists of a single 500-dimensional embedding layer, and we update the thresholds every 10 epochs. We tune other hyperparameters including Laplacian smoothing filter layers t, r^{st}_{pos}, r^{ed}_{pos}, r^{st}_{neg}, and r^{ed}_{neg} based on DBI. The detailed hyperparameter settings are reported in Table 4.2.

4.3.4   NODE CLUSTERING RESULTS

The node clustering results are presented in Table 4.3, where **bold** and <u>underlined</u> values indicate the highest scores in all methods and all baselines, respectively. Our observations are as follows.

AGE shows superior performance to baseline methods by a considerable margin, especially on Cora and Wiki datasets. Competing with the strongest baseline GALA, AGE outperforms it by 5.20% and 6.20% on Cora, by 18.45% and 13.11% on Wiki with respect to NMI and ARI. Such results show strong evidence advocating our proposed framework. For Citeseer and Pubmed, we give further analysis in Section 4.3.8.

Compared with GCN-based methods, AGE has simpler mechanisms than those in baselines, such as adversarial regularization or attention. The only trainable parameters are in the weight matrix of the 1-layer perceptron, which minimizes memory usage and improves training efficiency.

4.3.5   LINK PREDICTION RESULTS

In this section, we evaluate the quality of node embeddings on the link prediction task. Following the experimental settings of GALA, we conduct experiments on Cora and Citeseer, removing 5% edges for validation and 10% edges for test. The training procedure and hyper-parameters remain unchanged. Given the node embedding matrix $\mathbf{Z}$, we use a simple inner product decoder to get the predicted adjacency matrix

$$\hat{\mathbf{A}} = \sigma(\mathbf{Z}\mathbf{Z}^T), \tag{4.22}$$

where σ is the sigmoid function.

The experimental results are reported in Table 4.4. Compared with state-of-the-art unsupervised graph representation learning models, AGE outperforms them on both AUC and AP.

Table 4.3: Experimental results of node clustering

		Cora		Citeseer		Wiki		Pubmed	
Methods	Input	NMI	ARI	NMI	ARI	NMI	ARI	NMI	ARI
Kmeans	F	0.317	0.244	0.312	0.285	0.440	0.151	0.278	0.246
Spectral-F	F	0.147	0.071	0.203	0.183	0.464	0.254	0.309	0.277
Spectral-G	G	0.195	0.045	0.118	0.013	0.193	0.017	0.097	0.062
DeepWalk	G	0.327	0.243	0.089	0.092	0.324	0.173	0.102	0.088
TADW	F&G	0.441	0.332	0.291	0.228	0.271	0.045	0.244	0.217
GAE	F&G	0.482	0.302	0.221	0.191	0.345	0.189	0.249	0.246
VGAE	F&G	0.408	0.347	0.261	0.206	0.468	0.263	0.216	0.201
MGAE	F&G	0.489	0.436	0.416	0.425	0.510	0.379	0.282	0.248
ARGA	F&G	0.449	0.352	0.350	0.341	0.345	0.112	0.276	0.291
ARVGA	F&G	0.450	0.374	0.261	0.245	0.339	0.107	0.117	0.078
AGC	F&G	0.537	0.486	0.411	0.419	0.453	0.343	0.316	0.319
DAEGC	F&G	0.528	0.496	0.397	0.410	0.448	0.331	0.266	0.278
GALA	F&G	0.577	0.532	0.441	0.446	0.504	0.389	0.327	0.321
LS	F&G	0.493	0.373	0.419	0.433	0.534	0.317	0.300	0.315
LS+RA	F&G	0.580	0.545	0.410	0.403	0.566	0.382	0.291	0.301
LS+RX	F&G	0.479	0.423	0.416	0.424	0.543	0.365	0.285	0.251
AGE	F&G	**0.607**	**0.565**	**0.448**	**0.457**	**0.597**	**0.440**	0.316	**0.334**

Table 4.4: Experimental results of link prediction

	Cora		Citeseer	
Methods	AUC	AP	AUC	AP
GAE	0.910	0.920	0.895	0.899
VGAE	0.914	0.926	0.908	0.920
ARGA	0.924	0.932	0.919	0.930
ARVGA	0.924	0.926	0.924	0.930
GALA	0.921	0.922	0.944	0.948
AGE	**0.957**	**0.952**	**0.964**	**0.968**

It is worth noting that the training objectives of GAE/VGAE and ARGA/ARVGA are the adjacency matrix reconstruction loss. GALA also adds reconstruction loss for the link prediction task, while AGE does not utilize explicit links for supervision.

4.3.6   GAE VS. LS+RA

We use controlled experiments to verify hypothesis **H1**, evaluating the influence of entanglement of the filters and weight matrices. The compared methods are GAE and LS+RA, where the only difference between them is the position of the weight matrices. GAE, as we show in Fig. 4.2, combines the filter and weight matrix in each layer. LS+RA, however, moves weight matrices after the filter. Specifically, GAE has multiple GCN layers where each one contains a 64-dimensional linear layer, a ReLU activation layer and a graph convolutional filter. LS+RA stacks multiple graph convolutional filters and after which is a 1-layer 64-dimensional perceptron. Both embedding layers of the two models are 16-dimensional. Rest of the parameters are set to the same.

We report the NMI scores for node clustering on the four datasets with different number of filter layers in Fig. 4.4. The results show that LS+RA outperforms GAE under most circumstances with fewer parameters. Moreover, the performance of GAE decreases significantly as the filter layer increases, while LS+RA is relatively stable. A reasonable explanation to this phenomenon is stacking multiple graph convolution layers makes it harder to train all the weight matrices well. Also, the training efficiency will be affected by the deep network.

4.3.7   ABLATION STUDY

To validate **H2**, we first compare the four variants of AGE on the node clustering task. Our findings are listed below.

(1) Compared with raw features (Spectral-F), smoothed features (LS) integrate graph structure, thus perform better on node clustering. The improvement is considerable.

(2) The variants of our model, LS+RA and LS+RX, also show powerful performances compared with baseline methods, which results from our Laplacian smoothing filter. At the same time, AGE still outperforms the two variants, demonstrating that the adaptive optimization target is superior.

(3) Comparing the two reconstruction losses, reconstructing the adjacency matrix (LS+RA) performs better on Cora, Wiki, and Pubmed, while reconstructing the feature matrix (LS+RX) performs better on Citeseer. Such difference illustrates that structure information and feature information are of different importance across datasets, therefore either of them is not optimal universally. Furthermore, on Citeseer and Pubmed, the reconstruction losses contribute negatively to the smoothed features.

Then, we conduct ablation study on Cora to manifest the efficacy of four mechanisms in AGE. We set five variants of our model for comparison.

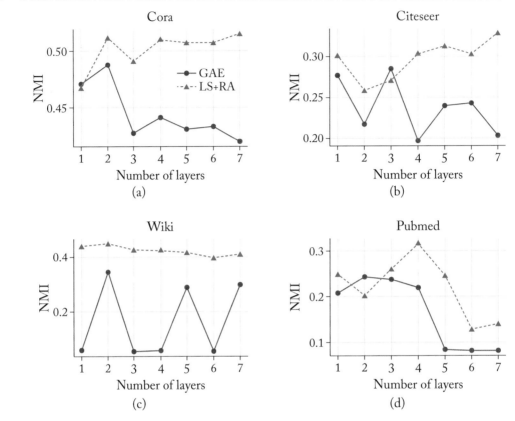

Figure 4.4: Controlled experiments comparing GAE and LS+RA.

All five variants cluster nodes by performing Spectral Clustering on the cosine similarity matrix of node features or embeddings. "Raw features" simply performs Spectral Clustering on raw node features; "+Filter" clusters nodes using smoothed node features; "+Encoder" initializes training set from the similarity matrix of smoothed node features, and learns node embeddings via the fixed training set; "+Adaptive" selects training samples adaptively with fixed thresholds; "+Thresholds Update" further adds thresholds update strategy and is exactly the full model.

In Table 4.5, it is obviously noticed that each part of our model contributes to the final performance, which evidently states the effectiveness of them. Additionally, we can observe that model supervised by the similarity of smoothed features ("+Encoder") outperforms almost all the baselines, giving verification to the rationality of our adaptive learning training objective.

Table 4.5: Ablation study

Model variants	Cora	
	NMI	ARI
Raw features	0.147	0.071
+Filter	0.493	0.373
+Encoder	0.558	0.521
+Adaptive	0.585	0.544
+Thresholds update	**0.607**	**0.565**

4.3.8   SELECTION OF k

As stated in previous sections, we select $k = 1/\lambda_{\max}$ while $\lambda_{\max}$ is the maximum eigenvalue of the renormalized Laplacian matrix. To verify the correctness of our hypothesis (**H3**), we first plot the eigenvalue distributions of the Laplacian matrix for benchmark datasets in Fig. 4.5. Then, we perform experiments with different k and the results are report in Fig. 4.6. From the two figures,we can make the following observations.

(1) The maximum eigenvalues of the four datasets are around 3/2, which supports our selection $k = 2/3$.

(2) In Fig. 4.6, it is clear that filters with $k = 2/3$ work best for Cora and Wiki datasets, since both NMI and ARI metrics reach the highest scores at $k = 2/3$. For Citeseer and Pubmed, there is little difference for various k.

(3) To further explain why some datasets are sensitive to k while some are not, we can look back into Fig. 4.5. Obviously, there are more high-frequency components in Cora and Wiki than Citeseer and Pubmed. Therefore, for Citeseer and Pubmed, filters with different k achieve similar effects.

Overall, for Laplacian smoothing filters, we can conclude that $k = 1/\lambda_{\max}$ is the optimal choice for Laplacian smoothing filters (**H3**).

4.3.9   VISUALIZATION

To intuitively show the learned node embeddings, we visualize the node representations in 2D space using t-SNE algorithm [Van Der Maaten, 2014]. The figures are shown in Fig. 4.7 and each subfigure corresponds to a variant in the ablation study. From the visualization, we can see that AGE can well cluster the nodes according to their corresponding classes. Additionally, as the model gets complete gradually, there are fewer overlapping areas and nodes belong to the same group gather together.

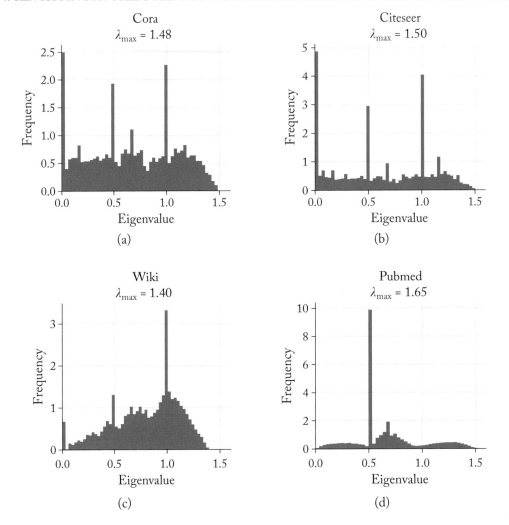

Figure 4.5: The eigenvalue distributions of benchmark datasets. $\lambda_{\max}$ is the maximum eigenvalue.

4.4   FURTHER READING

In this chapter, we revisit attributed NE from a GCN-based perspective and investigate the graph convolution operation in view of graph signal smoothing. We follow the conclusion [Li et al., 2018] that GCN is actually a Laplacian smoothing filter which removes high-frequency noises. Note that this conclusion is derived from a spectral-based analysis of graph neural networks (GNNs). There are also some works [Maron et al., 2019, Xu et al., 2018] discussing the expressive power of graph neural networks from the spatial view by Weisfeiler-Lehman (WL)

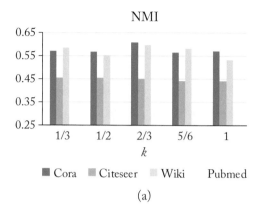

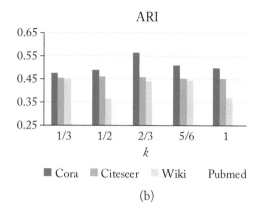

(a)   (b)

Figure 4.6: Influence of k on the two metrics.

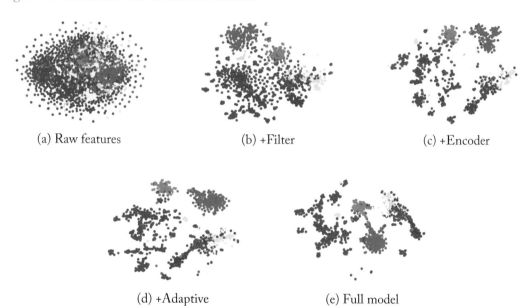

(a) Raw features   (b) +Filter   (c) +Encoder

(d) +Adaptive   (e) Full model

Figure 4.7: 2D visualization of node representations on Cora using t-SNE. The different colors represent different classes.

graph isomorphism test. More recently, researchers theoretically bridge the gap between spectral and spatial domains by a unified framework [Balcilar et al., 2020, Chen et al., 2020]. For more details about graph neural networks (e.g., what are spectral/spatial-based GNNs?), please refer to our survey [Zhou et al., 2018].

Part of this chapter was published in our KDD20 conference paper by Cui et al. [2020].

CHAPTER 5

Network Embedding for Graphs with Node Contents

In Chapter 3, we introduced TADW algorithm to learn NE from graph structure and node features jointly. In many real-world networks such as social network and citation network, nodes have rich text content which can be used to analyze their semantic aspects. In this chapter, we assume that a node usually shows different aspects when interacting with different neighbors (context), and thus should be assigned different embeddings. However, most existing models aim to learn a fixed context-free embedding for each node and neglect the diverse roles when interacting with other vertices. Therefore, we present Context-Aware Network Embedding (CANE), a novel NE model to address this issue. CANE learns context-aware embeddings with mutual attention mechanism and is expected to model the semantic relationships between vertices more precisely. In experiments, we compare CANE with baseline models on three real-world datasets. Experimental results show that CANE achieves significant improvement than previous NE methods on link prediction and comparable performance on node classification.

5.1   OVERVIEW

In real-world social networks, it is intuitive that a node has various aspects when interacting with different neighbors. For example, a researcher usually collaborates with various partners on diverse research topics (as illustrated in Fig. 5.1), a social-media user contacts with various friends sharing distinct interests, and a web page links to multiple pages for different purposes. However, most existing NE methods only assign one single embedding vector to each node, and give rise to the following two invertible issues: (1) These methods cannot flexibly cope with the aspect transition of a vertex when interacting with different neighbors. (2) In these models, a vertex tends to force the embeddings of its neighbors close to each other, which may not always be the case. For example, the left user and right user in Fig. 5.1, share less common interests, but are learned to be close to each other since they both link to the middle person. This will accordingly make node embeddings indiscriminative.

To address these issues, we introduce a **C**ontext-**A**ware **N**etwork **E**mbedding (CANE) framework for modeling relationships between vertices precisely. More specifically, we present CANE on information networks, where each node contains rich node content information. The significance of context is more critical for NE in this scenario. Without loss of generality, we

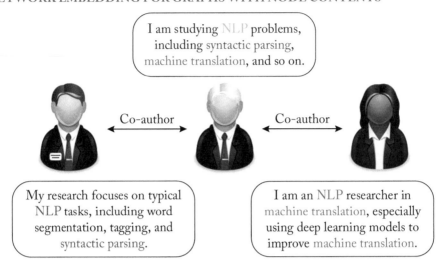

Figure 5.1: Example of a text-based information network. (Red, blue, and green fonts represent concerns of the left user, right user, and both, respectively.)

implement CANE on text-based information networks in this chapter, which can easily extend to other types of information networks.

In conventional NE models, each vertex is represented as a static embedding vector, denoted as **context-free embedding**. On the contrary, CANE assigns dynamic embeddings to a node according to different neighbors it interacts with, named as **context-aware embedding**. Take node u and its neighbo v for example. The context-free embedding of u remains unchanged when interacting with different neighbors. On the contrary, the context-aware embedding of u is dynamic when confronting different neighbors.

When u interacting with v, their context embeddings concerning each other are derived from their text information, S_u and S_v, respectively. For each vertex, we can easily use neural models, such as convolutional neural networks [Blunsom et al., 2014, Johnson and Zhang, 2014, Kim, 2014] and recurrent neural networks [Kiros et al., 2015, Tai et al., 2015], to build **context-free** text-based embedding. In order to realize **context-aware** text-based embeddings, we introduce the selective attention scheme and build **mutual attention** between u and v into these neural models. The mutual attention is expected to guide neural models to emphasize those words that are focused by its neighbor vertices and eventually obtain context-aware embeddings.

Both context-free embeddings and context-aware embeddings of each node can be efficiently learned together via concatenation using existing NE methods such as DeepWalk [Perozzi et al., 2014], LINE [Tang et al., 2015b], and node2vec [Grover and Leskovec, 2016].

We conduct experiments on three real-world datasets from different areas. Experimental results on link prediction reveal the effectiveness of our framework as compared to other state-of-the-art methods. The results suggest that context-aware embeddings are critical for network analysis, in particular for those tasks concerning about complicated interactions between vertices such as link prediction. We also explore the performance of our framework via node classification and case studies, which again confirms the flexibility and superiority of our models.

5.2   METHOD: CONTEXT-AWARE NETWORK EMBEDDING

5.2.1   PROBLEM FORMALIZATION

We first give basic notations and definitions used in this chapter. Suppose there is an information network $G = (V, E, T)$, where V is the set of vertices, $E \subseteq V \times V$ are edges between vertices, and T denotes the text information of vertices. Each edge $e_{u,v} \in E$ represents the relationship between two vertices (u, v), with an associated weight $w_{u,v}$. Here, the text information of a specific vertex $v \in V$ is represented as a word sequence $S_v = (w_1, w_2, \ldots, w_{n_v})$, where $n_v = |S_v|$. NRL aims to learn a low-dimensional embedding $\mathbf{v} \in \mathbb{R}^d$ for each vertex $v \in V$ according to its network structure and associated information, e.g., text and labels. Note that, $d \ll |V|$ is the dimension of representation space.

Definition 1. Context-free Embeddings: Conventional NRL models learn context-free embedding for each vertex. It means the embedding of a vertex is fixed and won't change with respect to its context information (i.e., another vertex it interacts with).

Definition 2. Context-aware Embeddings: Different from existing NRL models that learn context-free embeddings, CANE learns various embeddings for a vertex according to its different contexts. Specifically, for an edge $e_{u,v}$, CANE learns context-aware embeddings $\mathbf{v}_{(u)}$ and $\mathbf{u}_{(v)}$.

5.2.2   OVERALL FRAMEWORK

To take full use of both network structure and associated text information, we propose two types of embeddings for a vertex v, i.e., structure-based embedding $\mathbf{v}^s$ and text-based embedding $\mathbf{v}^t$. Structure-based embedding can capture the information in the network structure, while text-based embedding can capture the textual meanings lying in the associated text information. With these embeddings, we can simply concatenate them and obtain the vertex embeddings as $\mathbf{v} = \mathbf{v}^s \oplus \mathbf{v}^t$, where $\oplus$ indicates the concatenation operation. Note that the text-based embedding $\mathbf{v}^t$ can be either context-free or context-aware, which will be introduced later. When $\mathbf{v}^t$ is context-aware, the overall node embedding $\mathbf{v}$ will be context-aware as well.

With above definitions, CANE aims to maximize the overall objective of edges as follows:

$$\mathcal{L} = \sum_{e \in E} L(e). \tag{5.1}$$

Here, the objective of each edge $L(e)$ consists of two parts as follows:

$$L(e) = L_s(e) + L_t(e), \tag{5.2}$$

where $L_s(e)$ denotes the structure-based objective and $L_t(e)$ represents the text-based objective. Now we will introduce the two objectives, respectively.

5.2.3   STRUCTURE-BASED OBJECTIVE

Without loss of generality, we assume the network is directed, as an undirected edge can be considered as two directed edges with opposite directions and equal weights.

Thus, the structure-based objective aims to measure the log-likelihood of a directed edge using the structure-based embeddings as

$$L_s(e) = w_{u,v} \log p(\mathbf{v}^s | \mathbf{u}^s). \tag{5.3}$$

Following LINE [Tang et al., 2015b], we define the conditional probability of v generated by u in Eq. (5.3) as

$$p(\mathbf{v}^s | \mathbf{u}^s) = \frac{\exp(\mathbf{u}^s \cdot \mathbf{v}^s)}{\sum_{z \in V} \exp(\mathbf{u}^s \cdot \mathbf{z}^s)}. \tag{5.4}$$

5.2.4   TEXT-BASED OBJECTIVE

Nodes in real-world social networks usually accompany with associated text content information. Therefore, we propose the text-based objective to take advantage of these text information, as well as learn text-based embeddings for vertices.

The text-based objective $L_t(e)$ can be defined with various measurements. To be compatible with $L_s(e)$, we define $L_t(e)$ as follows:

$$L_t(e) = \alpha \cdot L_{tt}(e) + \beta \cdot L_{ts}(e) + \gamma \cdot L_{st}(e), \tag{5.5}$$

where α, β, and γ control the weights of various parts, and

$$\begin{aligned}
L_{tt}(e) &= w_{u,v} \log p(\mathbf{v}^t | \mathbf{u}^t), \\
L_{ts}(e) &= w_{u,v} \log p(\mathbf{v}^t | \mathbf{u}^s), \\
L_{st}(e) &= w_{u,v} \log p(\mathbf{v}^s | \mathbf{u}^t).
\end{aligned} \tag{5.6}$$

The conditional probabilities in Eq. (5.6) map the two types of vertex embeddings into the same representation space, but do not enforce them to be identical for the consideration of their own characteristics. Similarly, we employ softmax function for calculating the probabilities, as in Eq. (5.4).

The structure-based embeddings are regarded as parameters, the same as in conventional NE models. But for text-based embeddings, we intend to obtain them from associated text information of vertices. Besides, the text-based embeddings can be obtained either in context-free ways or context-aware ones. In the following sections, we will present both approaches, respectively.

5.2.5   CONTEXT-FREE TEXT EMBEDDING

There has been a variety of neural models to obtain text embeddings from a word sequence, such as convolutional neural networks (CNN) [Blunsom et al., 2014, Johnson and Zhang, 2014, Kim, 2014] and recurrent neural networks (RNN) [Kiros et al., 2015, Tai et al., 2015].

We investigate different neural networks for text modeling, including CNN, Bidirectional RNN, and GRU, and employ the best performed CNN, which can capture the local semantic dependency among words.

Taking the word sequence of a node as input, CNN obtains the text-based embedding through three layers, i.e., looking-up, convolution, and pooling.

Looking-up. Given a word sequence $S = (w_1, w_2, \ldots, w_n)$, the embedding looking-up layer transforms each word $w_i \in S$ into its corresponding word embedding $\mathbf{w}_i \in \mathbb{R}^{d'}$ and obtains embedding sequence as $\mathbf{S} = (\mathbf{w}_1, \mathbf{w}_2, \ldots, \mathbf{w}_n)$. Here, d' indicates the dimension of word embeddings.

Convolution. After looking-up, the convolution layer extracts local features of input embedding sequence $\mathbf{S}$. To be specific, it performs convolution operation over a sliding window of length l using a convolution matrix $\mathbf{C} \in \mathbb{R}^{d \times (l \times d')}$ as follows:

$$\mathbf{x}_i = \mathbf{C} \cdot \mathbf{S}_{i:i+l-1} + \mathbf{b}, \tag{5.7}$$

where $\mathbf{S}_{i:i+l-1}$ denotes the concatenation of word embeddings within the i-th window and $\mathbf{b}$ is the bias vector. Note that we add zero padding vectors at the end of the sentence.

Max-pooling. To obtain the text embedding $\mathbf{v}^t$, we operate max-pooling and nonlinear transformation over $\{\mathbf{x}_0^i, \ldots, \mathbf{x}_n^i\}$ as follows:

$$r_i = \tanh(\max(\mathbf{x}_0^i, \ldots, \mathbf{x}_n^i)). \tag{5.8}$$

At last, we encode the text information of a vertex with CNN and obtain its text-based embedding $\mathbf{v}^t = [r_1, \ldots, r_d]^T$. As $\mathbf{v}^t$ is irrelevant to the other vertices it interacts with, we name it as context-free text embedding.

5.2.6   CONTEXT-AWARE TEXT EMBEDDING

As stated before, we assume that a specific node plays different roles when interacting with others vertices. In other words, each vertex should have its own points of focus about a specific vertex, which leads to its context-aware text embeddings.

To achieve this, we employ **mutual attention** to obtain context-aware text embedding. The similar technique was originally applied in question answering as attentive pooling [dos Santos et al., 2016]. It enables the pooling layer in CNN to be aware of the vertex pair in an edge, in a way that text information from a vertex can directly affect the text embedding of the other vertex, and vice versa.

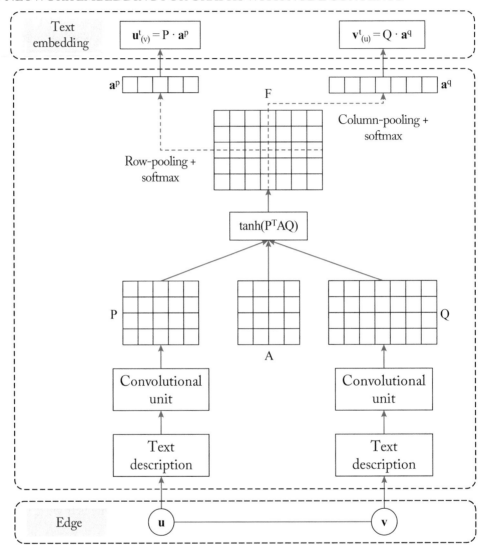

Figure 5.2: An illustration of context-aware text embedding.

In Fig. 5.2, we give an illustration of the generating process of context-aware text embedding. Given an edge $e_{u,v}$ with two corresponding text sequences S_u and S_v, we can get the matrices $\mathbf{P} \in \mathbb{R}^{d \times m}$ and $\mathbf{Q} \in \mathbb{R}^{d \times n}$ through convolution layer. Here, m and n represent the lengths of S_u and S_v, respectively. By introducing an attentive matrix $\mathbf{A} \in \mathbb{R}^{d \times d}$, we compute the correlation matrix $\mathbf{F} \in \mathbb{R}^{m \times n}$ as follows:

$$\mathbf{F} = \tanh(\mathbf{P}^T \mathbf{A} \mathbf{Q}). \tag{5.9}$$

Note that each element $\mathbf{F}_{i,j}$ in $\mathbf{F}$ represents the pair-wise correlation score between two hidden vectors, i.e., $\mathbf{P}_i$ and $\mathbf{Q}_j$.

After that, we conduct pooling operations along rows and columns of $\mathbf{F}$ to generate the importance vectors, named as row-pooling and column pooling, respectively. According to our experiments, mean-pooling performs better than max-pooling. Thus, we employ mean-pooling operation as follows:

$$
\begin{aligned}
g_i^p &= \mathbf{mean}(\mathbf{F}_{i,1}, \ldots, \mathbf{F}_{i,n}), \\
g_i^q &= \mathbf{mean}(\mathbf{F}_{1,i}, \ldots, \mathbf{F}_{m,i}).
\end{aligned}
\tag{5.10}
$$

The importance vectors of $\mathbf{P}$ and $\mathbf{Q}$ are obtained as $\mathbf{g}^p = [g_1^p, \ldots, g_m^p]^T$ and $\mathbf{g}^q = [g_1^q, \ldots, g_n^q]^T$.

Next, we employ softmax function to transform importance vectors $\mathbf{g}^p$ and $\mathbf{g}^q$ to attention vectors $\mathbf{a}^p$ and $\mathbf{a}^q$. For instance, the i-th element of $\mathbf{a}^p$ is formalized as follows:

$$
a_i^p = \frac{\exp(g_i^p)}{\sum_{j \in [1,m]} \exp(g_j^p)}.
\tag{5.11}
$$

At last, the context-aware text embeddings of u and v are computed as

$$
\begin{aligned}
\mathbf{u}_{(v)}^t &= \mathbf{P}\mathbf{a}^p, \\
\mathbf{v}_{(u)}^t &= \mathbf{Q}\mathbf{a}^q.
\end{aligned}
\tag{5.12}
$$

Now, given an edge (u, v), we can obtain the context-aware embeddings of vertices with their structure embeddings and context-aware text embeddings as $\mathbf{u}_{(v)} = \mathbf{u}^s \oplus \mathbf{u}_{(v)}^t$ and $\mathbf{v}_{(u)} = \mathbf{v}^s \oplus \mathbf{v}_{(u)}^t$.

5.2.7   OPTIMIZATION OF CANE

According to Eqs. (5.3) and (5.6), CANE aims to maximize several conditional probabilities between $\mathbf{u} \in \{\mathbf{u}^s, \mathbf{u}_{(v)}^t\}$ and $\mathbf{v} \in \{\mathbf{v}^s, \mathbf{v}_{(u)}^t\}$. It is intuitive that optimizing the conditional probability using softmax function is computationally expensive. Thus, we employ negative sampling [Mikolov et al., 2013b] and transform the objective into the following form:

$$
\log \sigma(\mathbf{u}^T \cdot \mathbf{v}) + \sum_{i=1}^{k} E_{z \sim P(v)}[\log \sigma(-\mathbf{u}^T \cdot \mathbf{z})],
\tag{5.13}
$$

where k is the number of negative samples and σ represents the sigmoid function. $P(v) \propto d_v^{3/4}$ denotes the distribution of vertices, where d_v is the out-degree of v.

Afterward, we employ Adam [Kingma and Ba, 2015] to optimize the transformed objective. Note that CANE is capable of zero-shot scenarios, by generating text embeddings of new vertices with well-trained CNN. CANE is effective for modeling the relationship between vertices and the learned context-aware embeddings can also compose high-quality context-free embeddings.

Table 5.1: Statistics of datasets

Datasets	Cora	HepTh	Zhihu
#Vertices	2,277	1,038	10,000
#Edges	5,214	1,990	43,894
#Labels	7	—	—

5.3 EMPIRICAL ANALYSIS

To investigate the effectiveness of CANE on modeling relationships between vertices, we conduct experiments of link prediction on several real-world datasets. Besides, we also employ node classification to verify whether context-aware embeddings of a vertex can compose a high-quality context-free embedding in return.

5.3.1 DATASETS

We select three real-world network datasets as follows.

Cora[1] is a typical paper citation network constructed by McCallum et al. [2000]. After filtering out papers without text information, there are 2,277 machine learning papers in this network, which are divided into 7 categories.

HepTh[2] (High Energy Physics Theory) is another citation network from arXiv[3] released by Leskovec et al. [2005]. We filter out papers without abstract information and retain 1,038 papers at last.

Zhihu[4] is the largest online Q&A website in China. Users follow each other and answer questions on this site. We randomly crawl 10,000 active users from Zhihu, and take the descriptions of their concerned topics as text information.

The detailed statistics are listed in Table 5.1.

5.3.2 BASELINES

We employ the following methods as baselines.

---

[1]https://people.cs.umass.edu/$\sim$mccallum/data.html
[2]https://snap.stanford.edu/data/cit-HepTh.html
[3]https://arxiv.org/
[4]https://www.zhihu.com/

Structure-only:

MMB [Airoldi et al., 2008] (Mixed Membership Stochastic Blockmodel) is a conventional graphical model of relational data. It allows each vertex to randomly select a different "topic" when forming an edge.

DeepWalk [Perozzi et al., 2014] performs random walks over networks and employ Skip-Gram model [Mikolov et al., 2013a] to learn vertex embeddings.

LINE [Tang et al., 2015b] learns vertex embeddings in large-scale networks using first-order and second-order proximities.

Node2vec [Grover and Leskovec, 2016] proposes a biased random walk algorithm based on DeepWalk to explore neighborhood architecture more efficiently.

Structure and Text:

Naive Combination: We simply concatenate the best-performed structure-based embeddings with CNN based embeddings to represent the vertices.

TADW [Yang et al., 2015] employs matrix factorization to incorporate text features of vertices into NEs.

CENE [Sun et al., 2016] leverages both structure and textural information by regarding text content as a special kind of vertices, and optimizes the probabilities of heterogeneous links.

5.3.3   EVALUATION METRICS AND EXPERIMENTAL SETTINGS

For link prediction, we adopt a standard evaluation metric **AUC**, which represents the probability that vertices in a random unobserved link are more similar than those in a random nonexistent link.

For vertex classification, we employ L2-regularized logistic regression (L2R-LR) [Fan et al., 2008] to train classifiers, and evaluate the classification accuracies of various methods.

To be fair, we set the embedding dimension to 200 for all methods. In LINE, we set the number of negative samples to 5; we learn the 100-dimensional first-order and second-order embeddings, respectively, and concatenate them to form the 200-dimensional embeddings. In node2vec, we employ grid search and select the best-performed hyper-parameters for training. We also apply grid search to set the hyper-parameters α, β, and γ in CANE. Besides, we set the number of negative samples k to 1 in CANE to speed up the training process. To demonstrate the effectiveness of considering attention mechanism and two types of objectives in Eqs. (5.3) and (5.6), we design three versions of CANE for evaluation, i.e., CANE with text only, CANE without attention, and CANE.

Table 5.2: AUC values on Cora ($\alpha = 1.0, \beta = 0.3, \gamma = 0.3$)

%Training edges	15%	25%	35%	45%	55%	65%	75%	85%	95%
MMB	54.7	57.1	59.5	61.9	64.9	67.8	71.1	72.6	75.9
DeepWalk	56.0	63.0	70.2	75.5	80.1	85.2	85.3	87.8	90.3
LINE	55.0	58.6	66.4	73.0	77.6	82.8	85.6	88.4	89.3
node2vec	55.9	62.4	66.1	75.0	78.7	81.6	85.9	87.3	88.2
Naive combination	72.7	82.0	84.9	87.0	88.7	91.9	92.4	93.9	94.0
TADW	86.6	88.2	90.2	90.8	90.0	93.0	91.0	93.4	92.7
CENE	72.1	86.5	84.6	88.1	89.4	89.2	93.9	95.0	95.9
CANE (text only)	78.0	80.5	83.9	86.3	89.3	91.4	91.8	91.4	93.3
CANE (w/o attention)	85.8	90.5	91.6	93.2	93.9	94.6	95.4	95.1	95.5
CANE	**86.8**	**91.5**	**92.2**	**93.9**	**94.6**	**94.9**	**95.6**	**96.6**	**97.7**

Table 5.3: AUC values on HepTh ($\alpha = 0.7, \beta = 0.2, \gamma = 0.2$)

%Training edges	15%	25%	35%	45%	55%	65%	75%	85%	95%
MMB	54.6	57.9	57.3	61.6	66.2	68.4	73.6	76.0	80.3
DeepWalk	55.2	66.0	70.0	75.7	81.3	83.3	87.6	88.9	88.0
LINE	53.7	60.4	66.5	73.9	78.5	83.8	87.5	87.7	87.6
node2vec	57.1	63.6	69.9	76.2	84.3	87.3	88.4	89.2	89.2
Naive combination	78.7	82.1	84.7	88.7	88.7	91.8	92.1	92.0	92.7
TADW	87.0	89.5	91.8	90.8	91.1	92.6	93.5	91.9	91.7
CENE	86.2	84.6	89.8	91.2	92.3	91.8	93.2	92.9	93.2
CANE (text only)	83.8	85.2	87.3	88.9	91.1	91.2	91.8	93.1	93.5
CANE (w/o attention)	84.5	89.3	89.2	91.6	91.1	91.8	92.3	92.5	93.6
CANE	**90.0**	**91.2**	**92.0**	**93.0**	**94.2**	**94.6**	**95.4**	**95.7**	**96.3**

5.3.4   LINK PREDICTION

As shown in Tables 5.2, 5.3, and 5.4, we evaluate the AUC values while removing different ratios of edges on Cora, HepTh, and Zhihu, respectively. Note that when we only keep 5% edges for training, most nodes are isolated, which results in the poor and meaningless performance of all the methods. Thus, we omit the results under this training ratio. From these tables, we have the following observations.

(1) CANE consistently achieves significant improvement comparing to all the baselines on all different datasets and different training ratios. It indicates the effectiveness of CANE

Table 5.4: AUC values on Zhihu ($\alpha = 1.0, \beta = 0.3, \gamma = 0.3$)

%Training edges	15%	25%	35%	45%	55%	65%	75%	85%	95%
MMB	51.0	51.5	53.7	58.6	61.6	66.1	68.8	68.9	72.4
DeepWalk	56.6	58.1	60.1	60.0	61.8	61.9	63.3	63.7	67.8
LINE	52.3	55.9	59.9	60.9	64.3	66.0	67.7	69.3	71.1
node2vec	54.2	57.1	57.3	58.3	58.7	62.5	66.2	67.6	68.5
Naive combination	55.1	56.7	58.9	62.6	64.4	68.7	68.9	69.0	71.5
TADW	52.3	54.2	55.6	57.3	60.8	62.4	65.2	63.8	69.0
CENE	56.2	57.4	60.3	63.0	66.3	66.0	70.2	69.8	73.8
CANE (text only)	55.6	56.9	57.3	61.6	63.6	67.0	68.5	70.4	73.5
CANE (w/o attention)	56.7	59.1	60.9	64.0	66.1	68.9	69.8	71.0	74.3
CANE	56.8	59.3	62.9	64.5	68.9	70.4	71.4	73.6	75.4

when applied to link prediction task, and verifies that CANE has the capability of modeling relationships between vertices precisely.

(2) What calls for special attention is that both CENE and TADW exhibit unstable performance under various training ratios. Specifically, CENE performs poorly under small training ratios, because it reserves much more parameters (e.g., convolution kernels and word embeddings) than TADW, which need more data for training. Different from CENE, TADW performs much better under small training ratios, because DeepWalk-based methods can explore the sparse network structure well through random walks even with limited edges. However, it achieves poor performance under large ones, as its simplicity and the limitation of bag-of-words assumption. On the contrary, CANE has a stable performance in various situations. It demonstrates the flexibility and robustness of CANE.

(3) By introducing attention mechanism, the learned context-aware embeddings obtain considerable improvements than the ones without attention. It verifies our assumption that a specific vertex should play different roles when interacting with other vertices, and thus benefits the relevant link prediction task.

To summarize, all the above observations demonstrate that CANE can learn high-quality context-aware embeddings, which can estimate the relationship between vertices precisely. Moreover, the experimental results on link prediction task state the effectiveness and robustness of CANE.

5.3.5   NODE CLASSIFICATION

In CANE, we obtain various embeddings of a vertex according to the vertex it connects to. It's intuitive that the obtained context-aware embeddings are naturally applicable to link prediction

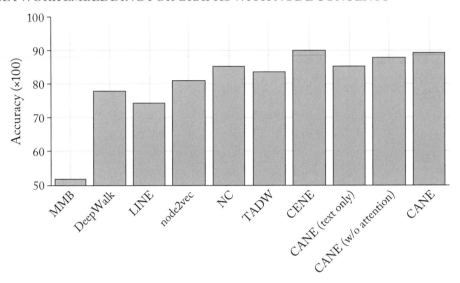

Figure 5.3: Node classification results on Cora.

task. However, network analysis tasks, such as vertex classification and clustering, require a global embedding, rather than several context-aware embeddings for each vertex.

To demonstrate the capability of CANE to solve these issues, we generate the global embedding of a vertex u by simply averaging all the context-aware embeddings as follows:

$$\mathbf{u} = \frac{1}{N} \sum_{(u,v)|(v,u)\in E} \mathbf{u}_{(v)},$$

where N indicates the number of context-aware embeddings of u.

With the generated global embeddings, we conduct two-fold cross-validation and report the average accuracy of vertex classification on Cora. As shown in Fig. 5.3, we observe that the following.

(1) CANE achieves comparable performance with state-of-the-art model CENE. It states that the learned context-aware embeddings can transform into high-quality context-free embeddings through simple average operation, which can be further employed to other network analysis tasks.

(2) With the introduction of mutual attention mechanism, CANE has an encouraging improvement than the one without attention, which is in accordance with the results of link prediction. It denotes that CANE is flexible to various network analysis tasks.

Edge #1: (A, B)

Machine Learning research making great progress many directions This article summarizes four directions discusses current open problems The four directions improving classification accuracy learning ensembles classifiers methods scaling supervised learning algorithms reinforcement learning learning complex stochastic models

The problem making optimal decisions uncertain conditions central Artificial Intelligence If state world known times world modeled Markov Decision Process MDP MDPs studied extensively many methods known determining optimal courses action policies The realistic case state information partially observable Partially Observable Markov Decision Processes POMDPs received much less attention The best exact algorithms problems inefficient space time We introduce Smooth Partially Observable Value Approximation SPOVA new approximation method quickly yield good approximations improve time This method combined reinforcement learning methods combination effective test cases

Edge #2: (A, C)

Machine Learning research making great progress many directions This article summarizes four directions discusses current open problems The four directions improving classification accuracy learning ensembles classifiers methods scaling supervised learning algorithms reinforcement learning learning complex stochastic models

In context machine learning examples paper deals problem estimating quality attributes without dependencies among Kira Rendell developed algorithm called RELIEF shown efficient estimating attributes Original RELIEF deal discrete continuous attributes limited twoclass problems In paper RELIEF analysed extended deal noisy incomplete multiclass data sets The extensions verified various artificial one well known realworld problem

Figure 5.4: Visualizations of mutual attention.

5.3.6   CASE STUDY

To demonstrate the effectiveness of mutual attention on selecting meaningful features from text information, we visualize the heat maps of two node pairs in Fig. 5.4. Note that every word in this figure accompanies with various background colors. The stronger the background color is, the larger the weight of this word is. The weight of each word is calculated according to the attention weights as follows.

For each vertex pair, we can get the attention weight of each convolution window according to Eq. (5.11). To obtain the weights of words, we assign the attention weight to each word in this window, and add the attention weights of a word together as its final weight.

The proposed attention mechanism makes the relations between vertices explicit and interpretable. We select three connected vertices in Cora dataset for example, denoted as A, B, and C. From Fig. 5.4, we observe that, though both paper B and C have citation relations with paper A, they concern about different parts of A. The attention weights over A in edge #1 are assigned to "reinforcement learning." On the contrary, the weights in edge #2 are assigned to "machine learning," "supervised learning algorithms," and "complex stochastic models." Moreover, all these key elements in A have corresponding words in B and C. It's intuitive that these key elements can serve as an explanation of the citation relations. This finding demonstrates the effectiveness of mutual attention mechanism, as well as the capability of CANE for modeling relations precisely.

5.4   FURTHER READING

The idea of mutual attention and context-aware modeling origins from natural language processing area [dos Santos et al., 2016, Rocktäschel et al., 2015]. To the best of our knowledge, NE models proposed before CANE all focused on learning context-free embeddings and ignored the diverse roles when a vertex interacts with others. In contrast, CANE assumes that a node has different embeddings according to which node it interacts with, and propose learn context-aware node embeddings. Follow-up work of CANE has investigated more fine-grained word-to-word matching mechanism [Shen et al., 2018], as well as context-aware user-item [Wu et al., 2019b] or music [Wang et al., 2020] embeddings for recommendation systems.

Note that there is also another trend of work [Epasto and Perozzi, 2019, Liu et al., 2019a, Yang et al., 2018b] which aims to learn multiple context-free embeddings for each node. Each embedding of a node is expected to represent a facet/aspect/community. There are two major differences between these methods and CANE.

(1) The context-aware embeddings learned by CANE is conditioned on each neighbor and thus more fine-grained. In contrast, the number of embeddings for each node in Epasto and Perozzi [2019], Liu et al. [2019a] and Yang et al. [2018b] is usually small and each embedding will encode more high-level information than a single edge.

(2) The parameters to be learned in CANE lie in the mutual attention mechanism, while those in Epasto and Perozzi [2019], Liu et al. [2019a] and Yang et al. [2018b] are the embedding itself.

In fact, the idea of learning multiple embeddings and forcing them to represent different aspects is also related to the term "disentangled representation" [Chen et al., 2016].

Part of this chapter was published in our ACL17 conference paper by Tu et al. [2017a].

CHAPTER 6

Network Embedding for Graphs with Node Labels

Aforementioned NE methods learned low-dimensional representations for nodes in networks in a fully unsupervised manner. However, the learned representations usually lack the ability of discrimination when applied to machine learning tasks, such as node classification. In this chapter, we overcome this challenge by proposing a novel semi-supervised model, max-margin Deep-Walk (MMDW). MMDW is a unified NRL framework that jointly optimizes the max-margin classifier and the aimed NE model. Influenced by the max-margin classifier, the learned representations not only contain the network structure, but also have the characteristic of discrimination. The visualizations of learned representations indicate that MMDW is more discriminative than unsupervised ones, and the experimental results on node classification demonstrate that MMDW achieves a significant improvement than previous NE methods.

6.1 OVERVIEW

Most of previous NE models are learned in unsupervised schemas. Though the learned representations can be applied to various tasks, they could be weak in particular prediction tasks. It is worth pointing out that there are many additional labeling information for network vertices in real world. For example, pages in Wikipedia have their categorical labels, like "arts," "history," "science;" papers in Cora and Citeseer are also stored with field labels for easy retrieval. Such labeling information usually contains a useful summarization or features of nodes, but is not directly utilized in previous network representation learning models.

It is nontrivial to integrate labeling information into network representation learning and learn discriminative representations for network nodes. Inspired by the max-margin principle, we propose MMDW, a discriminative NRL model, to seek predictive representations for nodes in a network.

As illustrated in Fig. 6.1, MMDW first learns DeepWalk as matrix factorization. Afterward, it trains a max-margin based classifier (e.g., support vector machine [Hearst et al., 1998]) and enlarges the distance between support vectors and classification boundaries. In other words, MMDW jointly optimizes the max-margin based classifier (e.g., support vector machine) and NRL model. Influenced by max-margin classifier, the representations of vertices are therefore more discriminative and more suitable for prediction tasks.

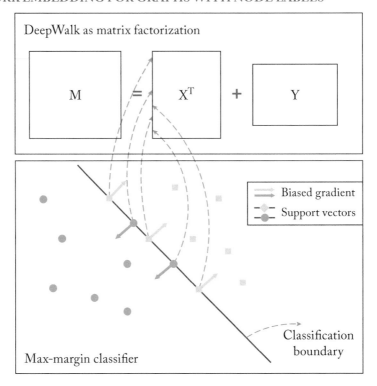

Figure 6.1: An illustration of Max-Margin DeepWalk.

We conduct node classification experiments on several real-world datasets to verify the effectiveness of MMDW. The experimental results demonstrate that MMDW significantly outperforms traditional NRL models. It achieves 5–10% improvements compared to typical NRL methods. Besides, we also compare the visualizations using t-SNE to illustrate the discrimination of MMDW.

6.2   METHOD: MAX-MARGIN DEEPWALK

In this section, we present a novel semi-supervised NE model, MMDW, that utilizes the labeling information when learning node representations. MMDW is a unified learning framework based on matrix factorization. In this model, we optimize the max-margin based classifier (SVM) as well as the aimed matrix factorization model. In contrast, the conventional approaches usually learn representations without leveraging the labeling information and apply the learned representations into classification tasks. The node labels are not able to influence the way representations are learned. Therefore, the learned representations are often found not discriminative enough.

6.2.1   PROBLEM FORMALIZATION

Suppose there is a network $G = (V, E)$, where V is the set of all vertices, and E are connections between these vertices, i.e., $E \subset V \times V$, NE aims to build a low-dimensional representation $\mathbf{x}_v \in \mathbb{R}^k$ for each vertex $v \in V$, where k is the dimension of representation space and expected much smaller than $|V|$. The learned representations encode semantic roles of vertices in the network, which can be used to measure relatedness between vertices, and can also play as features for classification tasks. With the corresponding label $l \in \{1, \cdots, m\}$, classifiers like logistic regression and SVM can be trained.

6.2.2   DEEPWALK BASED ON MATRIX FACTORIZATION

We incorporate the max-margin constraint into DeepWalk based on its matrix factorization form introduced in previous chapters.

Denote A as the transition matrix, which can be seen as the row normalized adjacency matrix, we follow the settings in TADW [Yang et al., 2015] and aim to factorize matrix $M = (A + A^2)/2$ to simulate DeepWalk algorithm. Specifically, we aim to find matrices $X \in \mathbb{R}^{k \times |V|}$ and $Y \in \mathbb{R}^{k \times |V|}$ as node embeddings to minimize

$$\min_{X,Y} \mathcal{L}_{DW} = \min_{X,Y} ||M - (X^T Y)||_2^2 + \frac{\lambda}{2}(||X||_2^2 + ||Y||_2^2), \qquad (6.1)$$

where the factor λ controls the weight of regularization part.

6.2.3   MAX-MARGIN DEEPWALK

Max-margin methods, such as SVMs [Hearst et al., 1998], are usually applied to various discriminative problems including document categorization and handwritten character recognition.

In this chapter, we take the learned representations X as features and train an SVM for node classification. Suppose the training set is $\mathcal{T} = \{(\mathbf{x}_1, l_1), \cdots, (\mathbf{x}_T, l_T)\}$, the multi-class SVM aims to find optimal linear functions by solving the following constrained optimization problem:

$$\min_{W,\boldsymbol{\xi}} \mathcal{L}_{SVM} = \min_{W,\boldsymbol{\xi}} \frac{1}{2} ||W||_2^2 + C \sum_{i=1}^{T} \xi_i$$

$$\text{s.t.} \quad \mathbf{w}_{l_i}^T \mathbf{x}_i - \mathbf{w}_j^T \mathbf{x}_i \geq e_i^j - \xi_i, \quad \forall i, j, \qquad (6.2)$$

where

$$e_i^j = \begin{cases} 1, & \text{if} \quad l_i \neq j, \\ 0, & \text{if} \quad l_i = j. \end{cases} \qquad (6.3)$$

Here, $W = [\mathbf{w}_1, \cdots, \mathbf{w}_m]^T$ is the weight matrix of SVM, and $\boldsymbol{\xi} = [\xi_1, \cdots, \xi_T]$ is the slack variable that tolerates errors in the training set.

As mentioned in previous parts, such pipeline method could not affect the way node representations are learned. With learned representations, SVM only helps to find optimal classifying boundaries. As a result, the representations themselves are not discriminative enough.

Inspired by max-margin learning on topic model [Zhu et al., 2012], we present max-margin DeepWalk MMDW to learn discriminative representations of vertices in a network. MMDW aims to optimize the max-margin classifier of SVM as well as matrix factorization based DeepWalk. The objective is defined as follows:

$$\min_{X,Y,W,\boldsymbol{\xi}} \mathcal{L} = \min_{X,Y,W,\boldsymbol{\xi}} \mathcal{L}_{DW} + \frac{1}{2}\|W\|_2^2 + C\sum_{i=1}^{T}\xi_i \tag{6.4}$$

$$\text{s.t.} \quad \mathbf{w}_{l_i}^T\mathbf{x}_i - \mathbf{w}_j^T\mathbf{x}_i \geq e_i^j - \xi_i, \quad \forall i, j.$$

6.2.4   OPTIMIZATION OF MMDW

The parameters in the objective Eq. (6.4) include vertex representation matrix X, context representation matrix Y, weight matrix W and slack variable vector $\boldsymbol{\xi}$. We employ an effective optimization strategy by optimizing the two parts separately. With the introduction of Biased Gradient, the matrix factorization is significantly affected by trained max-margin classifier. We perform our optimization algorithm as follows.

Optimization over W and $\boldsymbol{\xi}$:

When $\mathbf{X}$ and $\mathbf{Y}$ are fixed, the primal problem Eq. (6.4) becomes the same as a standard multi-class SVM problem, which has the following dual form:

$$\min_{\mathbf{Z}} \frac{1}{2}\|W\|_2^2 + \sum_{i=1}^{T}\sum_{j=1}^{m}e_i^j z_i^j \tag{6.5}$$

$$\text{s.t.} \quad \sum_{j=1}^{m} z_i^j = 0, \quad \forall i$$

$$z_i^j \leq C_{l_i}^j, \quad \forall i, j,$$

where

$$\mathbf{w}_j = \sum_{i=1}^{l} z_i^j \mathbf{x}_i, \quad \forall j$$

and

$$C_{y_i}^m = \begin{cases} 0, & \text{if} \quad y_i \neq m, \\ C, & \text{if} \quad y_i = m. \end{cases}$$

Here, the lagrangian multiplier α_i^j is replaced by $C_{l_i}^j - z_i^j$ for short.

To solve this dual problem, we utilize a coordinate descent method to decompose Z into blocks $[\mathbf{z}_1, \cdots, \mathbf{z}_T]$, where

$$\mathbf{z}_i = [z_i^1, \cdots, z_i^m]^T, \quad i = 1, \cdots, T.$$

An efficient sequential dual method [Keerthi et al., 2008] is applied to solve the sub-problem formed by $\mathbf{z}_i$.

Optimization over X and Y:

When W and $\boldsymbol{\xi}$ are fixed, the primal problem Eq. (6.4) turns into minimizing the square loss of matrix factorization with additional boundary constraints as follows:

$$\min_{X,Y} \mathcal{L}_{DW}(X, Y; M, \lambda) \tag{6.6}$$
$$\text{s.t.} \quad \mathbf{w}_{l_i}^T \mathbf{x}_i - \mathbf{w}_j^T \mathbf{x}_i \geq e_i^j - \xi_i, \quad \forall i, j.$$

Without the consideration of constraints, we have

$$\frac{\partial \mathcal{L}}{\partial X} = \lambda X - Y(M - X^T Y),$$
$$\frac{\partial \mathcal{L}}{\partial Y} = \lambda Y - X(M - X^T Y). \tag{6.7}$$

$\forall i \in \mathcal{T}, j \in 1, \ldots, m$, if $l_i \neq j$ and $\alpha_i^j \neq 0$, according to Karush–Kuhn–Tucker (KKT) conditions, we have

$$\mathbf{w}_{l_i}^T \mathbf{x}_i - \mathbf{w}_j^T \mathbf{x}_i = e_i^j - \xi_i. \tag{6.8}$$

When the decision boundary is fixed, we want to bias such support vector $\mathbf{x}_i$ toward the direction that favors a more accurate prediction. Thus, the biased vector can enlarge the discrimination.

Here we explain how the bias is calculated. Given a vertex $i \in \mathcal{T}$, for the j-th constraint, we add a component $\alpha_i^j(\mathbf{w}_{l_i} - \mathbf{w}_j)^T$ to $\mathbf{x}_i$, then the constraint becomes

$$(\mathbf{w}_{l_i} - \mathbf{w}_j)^T(\mathbf{x}_i + \alpha_i^j(\mathbf{w}_{l_i} - \mathbf{w}_j)) \tag{6.9}$$
$$= (\mathbf{w}_{l_i} - \mathbf{w}_j)^T \mathbf{x}_i + \alpha_i^j \|(\mathbf{w}_{l_i} - \mathbf{w}_j)\|_2^2$$
$$> e_i^j - \xi_i.$$

Note that we utilize the lagrangian multiplier α_i^j to judge whether the vector is on the decision boundary. Only $\mathbf{x}_i$ with $\alpha_i^j \neq 0$ is added a bias based on the j-th constraint.

For a vertex $i \in \mathcal{T}$, the gradient becomes $\frac{\partial \mathcal{L}}{\partial \mathbf{x}_i} + \eta \sum_{j=1}^m \alpha_i^j(\mathbf{w}_{l_i} - \mathbf{w}_j)^T$, which is named Biased Gradient. Here, η balances the primal gradient and the bias.

Before X is updated, W and $\boldsymbol{\xi}$ satisfy the KKT conditions of SVM, and this solution is initially optimal. But after updating X, the KKT conditions do not hold. This will lead to a slight increase of the objective, but this increase is usually within an acceptable range according to our experiments.

6.3   EMPIRICAL ANALYSIS

In this section, we conduct node classification to evaluate MMDW. Besides, we also visualize the learned representations to verify that MMDW is able to learn discriminative representations.

6.3.1   DATASETS AND EXPERIMENT SETTINGS

We employ the following three typical datasets for node classification.

Cora. Cora[1] is a research paper set constructed by McCallum et al. [2000]. It contains 2,708 machine learning papers which are categorized into 7 classes. The citation relationships between them form a typical network.

Citeseer. Citeseer is another research paper set constructed by McCallum et al. [2000]. It contains 3,312 publications and 4,732 connections between them. These papers are from six classes.

Wiki. Wiki [Sen et al., 2008] contains 2,405 web pages from 19 categories and 17,981 links between them. It's much denser than Cora and Citeseer.

For evaluation, we randomly sample a portion of labeled vertices and take their representations as features for training, and the rest are used for testing. We increase the training ratio from 10% to 90%, and employ multi-class SVM [Crammer and Singer, 2002] to build classifiers.

6.3.2   BASELINE METHODS

DeepWalk. DeepWalk [Perozzi et al., 2014] is a typical NRL model that learns node representations based on network structures. We set parameters in DeepWalk as follows: window size $K = 5$, walks per vertex $\gamma = 80$, and representation dimension $k = 200$. For a vertex v, we take the representation $\mathbf{v}$ as network feature vector.

DeepWalk as Matrix Factorization. In Chapter 3, we have introduced that DeepWalk can be trained in a matrix factorization form. Thus, we factorize the matrix $M = (A + A^2)/2$ and take the factorized matrix X as representations of vertices.

2nd-LINE. LINE [Tang et al., 2015b] is another NRL model that learns network representations in large-scale networks. We employ the second-order proximity LINE (2nd-LINE) to learn representations for directed networks. Same as DeepWalk, we also set the representation length as 200.

Table 6.1: Accuracy (%) of node classification on Cora

%Labeled nodes	10%	20%	30%	40%	50%	60%	70%	80%	90%
DW	68.51	73.73	76.87	78.64	81.35	82.47	84.31	85.58	85.61
MFDW	71.43	76.91	78.20	80.28	81.35	82.47	84.44	83.33	87.09
LINE	65.13	70.17	72.2	72.92	73.45	75.67	75.25	76.78	79.34
MMDW($\eta = 10^{-2}$)	**74.94**	**80.83**	**82.83**	**83.68**	**84.71**	**85.51**	**87.01**	**87.27**	**88.19**
MMDW($\eta = 10^{-3}$)	74.20	79.92	81.13	82.29	83.83	84.62	86.03	85.96	87.82
MMDW($\eta = 10^{-4}$)	73.66	79.15	80.12	81.31	82.52	83.90	85.54	85.95	87.82

Table 6.2: Accuracy (%) of node classification on Citeseer

%Labeled nodes	10%	20%	30%	40%	50%	60%	70%	80%	90%
DW	49.09	55.96	60.65	63.97	65.42	67.29	66.80	66.82	63.91
MFDW	50.54	54.47	57.02	57.19	58.60	59.18	59.17	59.03	55.35
LINE	39.82	46.83	49.02	50.65	53.77	54.2	53.87	54.67	53.82
MMDW($\eta = 10^{-2}$)	**55.60**	60.97	63.18	65.08	**66.93**	**69.52**	**70.47**	**70.87**	**70.95**
MMDW($\eta = 10^{-3}$)	55.56	**61.54**	**63.36**	**65.18**	66.45	69.37	68.84	70.25	69.73
MMDW($\eta = 10^{-4}$)	54.52	58.49	59.25	60.70	61.62	61.78	63.24	61.84	60.25

Table 6.3: Accuracy (%) of node classification on Wiki

%Labeled nodes	10%	20%	30%	40%	50%	60%	70%	80%	90%
DW	52.03	54.62	59.80	60.29	61.26	65.41	65.84	66.53	68.16
MFDW	56.40	60.28	61.90	63.39	62.59	62.87	64.45	62.71	61.63
LINE	52.17	53.62	57.81	57.26	58.94	62.46	62.24	66.74	67.35
MMDW($\eta = 10^{-2}$)	**57.76**	**62.34**	**65.76**	**67.31**	**67.33**	**68.97**	**70.12**	**72.82**	**74.29**
MMDW($\eta = 10^{-3}$)	54.31	58.69	61.24	62.63	63.18	63.58	65.28	64.83	64.08
MMDW($\eta = 10^{-4}$)	53.98	57.48	60.10	61.94	62.18	62.36	63.21	62.29	63.67

6.3.3   EXPERIMENTAL RESULTS AND ANALYSIS

Tables 6.1, 6.2, and 6.3 show classification accuracies with different training ratios on different datasets. In these tables, we denote DeepWalk as DW, matrix factorization form of DeepWalk

---

[1]https://people.cs.umass.edu/~mccallum/data.html

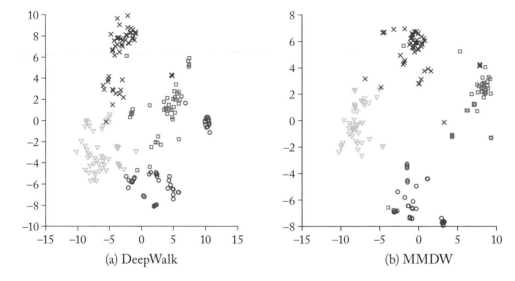

Figure 6.2: t-SNE 2D representations on Wiki (left: DeepWalk, right: MMDW).

as MFDW, and 2nd-LINE as LINE for short. We also show the performance of MMDW with η ranging from 10^{-4} to 10^{-2}. From these tables, we have the following observations.

(1) The proposed max-margin DeepWalk consistently and significantly outperforms all the baselines on all different datasets and different training ratios. Note that, MMDW achieves nearly 10% improvement on Citerseer and 5% improvement on Wiki when the training ratio is around 50%. DeepWalk cannot represent vertices in Citeseer and Wiki well, while MMDW is capable of managing such situation. These improvements demonstrate that MMDW is more robust and especially performs better when the quality of network representations is poor.

(2) What calls for special attention is that MMDW outperforms the original DeepWalk with half less training data on Citeseer and Wiki. It demonstrates that MMDW is effective when applied to predictive tasks.

(3) In contrast, the performance of original DeepWalk and DeepWalk as matrix factorization is unstable on various datasets. This indicates that the introduction of supervised information is important and MMDW is flexible to diversified networks.

Observations above demonstrate that MMDW is able to incorporate labeling information for generating high quality representations. MMDW is not application-specific. The learned representations of vertices can be utilized in many tasks, including vertex similarity, link prediction, classification and so on. The idea of biased gradient can be easily extended to other matrix factorization methods.

6.3.4   VISUALIZATION

To verify whether the learned representations is discriminative, we show the 2D representations of vertices on Wiki dataset using t-SNE visualization tool in Fig. 6.2. In this figure, each dot represents a vertex and each color represents a category. We randomly select four categories to show the trend more clearly.

From Fig. 6.2, we observe that MMDW learns a better clustering and separation of the vertices. On the contrary, the representations learned by DeepWalk tend to mix together. A well-separated representation is more discriminative and easier to categorize. These significant improvements prove the effectiveness of our discriminative learning model.

6.4   FURTHER READING

There have been some max-margin based learning methods in other fields. Roller [2004] first introduced max-margin principle into markov networks. Zhu et al. [2012] proposed maximum entropy discrimination LDA (MedLDA) to learn a discriminative topic model (e.g., latent Dirichlet allocation [Blei et al., 2003]). Besides, max-margin also benefits many NLP tasks, such as semantic parsing [Taskar et al., 2004] and word segmentation [Pei et al., 2014].

To the best of our knowledge, there is little work learning discriminative network representations with labeling information considered before MMDW. Most of the mentioned NRL methods are learned in an unsupervised manner. To fill this gap, we propose max-margin Deep-Walk (MMDW) to learn discriminative node representations. Some concurrent work [Li et al., 2016, Yang et al., 2016] of MMDW also explore the semi-supervised settings and yield competitive results. In fact, the semi-supervised setting also facilitates the rising of GCN [Kipf and Welling, 2017] for graph-based machine learning.

Part of this chapter was published in our IJCAI16 conference paper by Tu et al. [2016b].

PART III

Network Embedding with Different Characteristics

CHAPTER 7

Network Embedding for Community-Structured Graphs

Most existing NE methods focus on learning representations from local context of nodes (such as their neighbors) without assuming any global properties of the graph. Nevertheless, vertices in many complex networks also exhibit significant global patterns widely known as communities. In community-structured graphs, nodes in the same community tend to connect densely, and share common attributes. These patterns are expected to improve NE and benefit relevant evaluation tasks, such as link prediction and node classification. In this chapter, we present a novel NE model for community-structured graphs, named as Community-enhanced Network Representation Learning (CNRL). CNRL simultaneously detects community distribution of each vertex and learns embeddings of both vertices and communities. Moreover, the proposed community enhancement mechanism can be applied to various existing NE models. In experiments, we evaluate our model on node classification, link prediction, and community detection using several real-world datasets to demonstrate the effectiveness of CNRL.

7.1 OVERVIEW

Most NE methods learn node representations according to their *local context* information. For example, DeepWalk [Perozzi et al., 2014] performs random walks over the network topology and learns node representations by maximizing the likelihood of predicting their local contextual vertices in walk sequences; LINE [Tang et al., 2015b] learns vertex representations by maximizing the likelihood of predicting their neighbor vertices. Both *contextual vertices* in DeepWalk and *neighbor vertices* in LINE are local context.

In many scenarios, vertices will group into multiple communities with each community densely connected inside [Newman, 2006], which forms a *community-structured graph*. Vertices in a community usually share certain common attributes. For example, Facebook users with the same education-based attributes ("School name" or "Major") tend to form communities [Yang et al., 2013]. Hence, the community structure is an important *global pattern* of vertices, which is expected to benefit NRL as well as network analysis tasks.

Inspired by this, we propose a novel NRL model, community-enhanced NRL (CNRL), for community-structured graphs. It is worth pointing out that CNRL is highly motivated by the connection between text modeling and network modeling as well as DeepWalk [Perozzi et al., 2014]. As the analogy between words in text and vertices in walk sequences has been verified by DeepWalk [Perozzi et al., 2014], we assume that there are correlations between word preference on topics and vertex preference on communities as well. Although community information has been explored by some NRL models (e.g., MNMF [Wang et al., 2017g] and ComE [Cavallari et al., 2017]), we employ an entirely different approach to solve this problem by leveraging the analogy between topics and communities. This analogy is explicit and makes CNRL easy to be integrated into random-walk based NRL models.

The basic idea of CNRL is illustrated in Fig. 7.1. We consider each vertex is grouped into multiple communities, and these communities are overlapping. Different from conventional NRL methods, where node embedding is learned from local context nodes, CNRL will learn node embedding from both local context and global community information.

In CNRL, it is crucial to determine which community each node belongs to. Following the idea of topic models, each vertex in a specific sequence is assigned with a specific community, according to the community distribution of the vertex and that of the sequence. Afterward, each vertex and its assigned community are applied to predict its context vertices in the walk sequence. Therefore, representations of both vertices and communities are learned by maximizing the prediction likelihood. Note that community distributions of vertices are also updated iteratively in the representation learning process.

We implement CNRL on two typical random walk-based NRL models, DeepWalk [Perozzi et al., 2014] and node2vec [Grover and Leskovec, 2016], and conduct experiments on several real-world network datasets using the tasks of node classification, link prediction and community detection. Experimental results show that, CNRL can significantly improve the performance of all the tasks, and the superiority is consistent with respect to various datasets and training ratios. It demonstrates the effectiveness of considering global community information for network representation learning.

7.2   METHOD: COMMUNITY-ENHANCED NRL

We start with the necessary notations and formalizations of NRL for community-structured graphs.

7.2.1   PROBLEM FORMALIZATION

We denote a network as $G = (V, E)$, where V is the set of vertices and $E \subseteq (V \times V)$ is the set of edges, with $(v_i, v_j) \in E$ indicating there is an edge between v_i and v_j. For each vertex v, NRL aims to learn a low-dimensional vector denoted as $\mathbf{v} \in \mathbb{R}^d$. Here d represents the dimension of representation space.

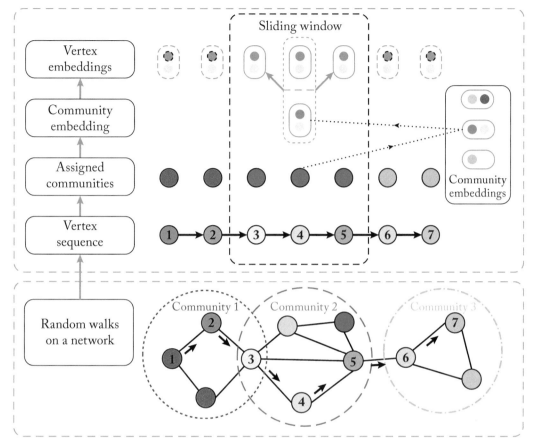

Figure 7.1: An illustration of CNRL.

The vertices in G can be grouped into K communities $C = \{c_1, \ldots, c_K\}$. The communities are usually overlapping. That is, one vertex may be the member of multiple communities in different degrees. Hence, we record the membership degree of a vertex v to a community c as the probability $\Pr(c|v)$, and the role of the vertex in c as the probability $\Pr(v|c)$. We will also learn representations of each community c, denoted as $\mathbf{c}$.

We will first give a brief review of DeepWalk and then implement the idea of CNRL by extending DeepWalk to community-enhanced DeepWalk. As the only difference between node2vec and DeepWalk is the generation methods of node sequences, which will not affect the extension of CNRL, we omit its implementation details.

7.2.2   DEEPWALK

DeepWalk performs random walks over the given network G first, and forms a set of walk sequences $S = \{s_1, \ldots, s_N\}$, where each sequence can be denoted as $s = \{v_1, \ldots, v_{|s|}\}$.

DeepWalk treats each walk sequence s as a word sequence by regarding vertices as words. By introducing Skip-Gram [Mikolov et al., 2013a], a widely used word representation learning algorithm, DeepWalk is able to learn vertex representations from the sequence set S.

More specifically, given a vertex sequence $s = \{v_1, \ldots, v_{|s|}\}$, each vertex v_i has $\{v_{i-t}, \ldots, v_{i+t}\} \setminus \{v_i\}$ as its local context vertices. Following Skip-Gram, DeepWalk learns node representations by maximizing the average log probability of predicting context nodes:

$$\mathcal{L}(s) = \frac{1}{|s|} \sum_{i=1}^{|s|} \sum_{i-t \leq j \leq i+t, j \neq i} \log \Pr(v_j | v_i), \tag{7.1}$$

where v_j is the context vertex of the vertex v_i, and the probability $\Pr(v_j | v_i)$ is defined by softmax function:

$$\Pr(v_j | v_i) = \frac{\exp(\mathbf{v}'_j \cdot \mathbf{v}_i)}{\sum_{v \in V} \exp(\mathbf{v}' \cdot \mathbf{v}_i)}. \tag{7.2}$$

Note that, as in Skip-Gram, each vertex v in DeepWalk also has two representation vectors, i.e., $\mathbf{v}_i$ when it is the center vertex and $\mathbf{v}'$ when it is the context vertex.

7.2.3   COMMUNITY-ENHANCED DEEPWALK

With random walk sequences, DeepWalk aims to maximize the local conditional probability of two vertices within a context window. That means, the co-occurrence of vertices in a sequence only relies on the affinity between vertices, while ignoring their global patterns. A critical global pattern of social networks is homophily, i.e., "birds of a feather flock together" [McPherson et al., 2001]. That is, those similar vertices sharing the same "feather" may group into communities. It's worth pointing out that such global pattern has not been considered by DeepWalk.

Communities provide rich contextual information of vertices. In order to take community information into consideration to provide richer context for NRL, we make two assumptions on the correlations among vertices, walk sequences and communities.

Assumption 1: Each vertex in a network may belong to multiple communities with different preferences, i.e., $\Pr(c|v)$, and each vertex sequence also owns its community distribution $\Pr(c|s)$.

We make another assumption about the particular community of a vertex in a sequence.

Assumption 2: A vertex in a specific walk sequence belongs to a distinct community, and the community is determined by the sequence's distribution over communities $\Pr(c|s)$ and the community's distribution over vertices $\Pr(v|c)$.

With the above assumptions and generated random walk sequences, we conduct the following two steps iteratively to detect community structures and learn representations of vertices and communities: (1) **Community Assignment.** We assign a discrete community label for each vertex in a particular walk sequence, according to both local context and global community distribution. (2) **Representation Learning.** Given a vertex and its community label, we learn optimized representations to maximize the log probability of predicting context vertices.

The two steps are demonstrated in Fig. 7.1. As shown in Fig. 7.1, we aim to learn an embedding for each vertex and each community. Besides, we also want to learn the community distribution of each vertex. We introduce the two steps in detail as follows.

Community Assignment

For a node v in a walk sequence s, we calculate the conditional probability of a community c as follows:

$$\Pr(c|v,s) = \frac{\Pr(c,v,s)}{\Pr(v,s)} \propto \Pr(c,v,s). \tag{7.3}$$

According to our assumptions, the joint distribution of (c,v,s) can be formalized as

$$\Pr(c,v,s) = \Pr(s)\Pr(c|s)\Pr(v|c), \tag{7.4}$$

where $\Pr(v|c)$ indicates the role of v in the community c, and $\Pr(c|s)$ indicates the local affinity of the sequence s with the community c. From Eqs. (7.3) and (7.4), we have

$$\Pr(c|v,s) \propto \Pr(v|c)\Pr(c|s). \tag{7.5}$$

In this chapter, we propose the following two strategies to implement $\Pr(c|v,s)$:

Statistic-based assignment. Following the Gibbs Sampling method of Latent Dirichlet Allocation (LDA) [Griffiths and Steyvers, 2004], we can estimate the conditional distributions of $\Pr(v|c)$ and $\Pr(c|s)$ as follows:

$$\Pr(v|c) = \frac{N(v,c) + \beta}{\sum_{\tilde{v} \in V} N(\tilde{v},c) + |V|\beta}, \tag{7.6}$$

$$\Pr(c|s) = \frac{N(c,s) + \alpha}{\sum_{\tilde{c} \in C} N(\tilde{c},s) + |K|\alpha}. \tag{7.7}$$

Here $N(v,c)$ indicates how many times the vertex v is assigned to the community c, and $N(c,s)$ indicates how many vertices in the sequence s are assigned to the community c. Both $N(v,c)$ and $N(c,s)$ will be updated dynamically as community assignments change. Moreover, α and β are smoothing factors following [Griffiths and Steyvers, 2004].

Embedding-based assignment. As CNRL will obtain the embeddings of vertices and communities, we can measure the conditional probabilities from an embedding view instead of global statistics. Hence, we formalize $\Pr(c|s)$ as follows:

$$\Pr(c|s) = \frac{\exp(\mathbf{c} \cdot \mathbf{s})}{\sum_{\tilde{c} \in C} \exp(\tilde{\mathbf{c}} \cdot \mathbf{s})}, \tag{7.8}$$

where $\mathbf{c}$ is the community vector learned by CNRL, and $\mathbf{s}$ is the semantic vector of the sequence s, which is the average of the embeddings of all vertices in s.

In fact, we can also calculate $\Pr(v|c)$ in the similar way:

$$\Pr(v|c) = \frac{\exp(\mathbf{v} \cdot \mathbf{c})}{\sum_{\tilde{v} \in V} \exp(\tilde{\mathbf{v}} \cdot \mathbf{c})}. \tag{7.9}$$

However, the usage of Eq. (7.9) will badly degrade the performance. We suppose the reason is that vertex embedding is not exclusively learned for measuring community membership. Hence, Eq. (7.9) could not be as discriminative as compared to the statistic-based Eq. (7.6). Therefore, in embedding-based assignment, we only calculate $\Pr(c|s)$ using embeddings and still use statistic-based $\Pr(v|c)$.

With estimated $\Pr(v|c)$ and $\Pr(c|s)$, we assign a discrete community label c for each vertex v in sequence s according to Eq. (7.5).

Representation Learning of Vertices and Communities

Given a certain vertex sequence $s = \{v_1, \ldots, v_{|s|}\}$, for each vertex v_i and its assigned community c_i, we will learn representations of both vertices and communities by maximizing the log probability of predicting context vertices using both v_i and c_i, which is formalized as follows:

$$\mathcal{L}(s) = \frac{1}{|s|} \sum_{i=1}^{|s|} \sum_{i-t \leq j \leq i+t, j \neq i} \log \Pr(v_j|v_i) + \log \Pr(v_j|c_i), \tag{7.10}$$

where $\Pr(v_j|v_i)$ is identical to Eq. (7.2), and $\Pr(v_j|c_i)$ is calculated similar to $\Pr(v_j|v_i)$ using a softmax function:

$$\Pr(v_j|c_i) = \frac{\exp(\mathbf{v}'_j \cdot \mathbf{c}_i)}{\sum_{\tilde{v} \in V} \exp(\tilde{\mathbf{v}}' \cdot \mathbf{c}_i)}. \tag{7.11}$$

Enhanced Vertex Representation

After the above-mentioned representation learning process, we will obtain representations of both vertices and communities, as well as community distribution of vertex, i.e., $\Pr(c|v)$. We can apply these results to build enhanced representation for each vertex, denoted as $\hat{\mathbf{v}}$. The enhanced representations encode both local and global context of vertices, which are expected to promote discriminability of network representation.

Specifically, $\hat{\mathbf{v}}$ consists of two parts, the original vertex representation $\mathbf{v}$ and its community representation $\mathbf{v}_c$, where

$$\mathbf{v}_c = \sum_{\tilde{c} \in C} \Pr(\tilde{c}|v) \tilde{\mathbf{c}}. \tag{7.12}$$

Table 7.1: Statistics of the real-world networks

Datasets	Cora	Citeseer	Wiki	BlogCatalog
# Vertices	2,708	3,312	2,405	10,312
# Edges	5,429	4,732	15,985	333,983
# Labels	7	6	19	47
Avg. degree	4.01	2.86	6.65	32.39

Afterward, we concatenate these two parts and obtain $\hat{\mathbf{v}} = \mathbf{v} \oplus \mathbf{v}_c$.

In experiments, we will investigate the performance of the enhanced vertex representations on network analysis tasks.

7.3   EMPIRICAL ANALYSIS

We adopt the tasks of node classification and link prediction to evaluate the performance of node embeddings. Besides, we also investigate the effectiveness of our model for community detection.

7.3.1   DATASETS

We conduct experiments on four widely adopted network datasets, including Cora, Citeseer, Wiki, and BlogCatalog. Cora and Citeseer [McCallum et al., 2000] are both research paper sets, where the citation relationship between papers forms social networks. Wiki [Sen et al., 2008] is a Web page collection from Wikipedia, and the hyperlinks among these pages compose a Web network. BlogCatalog [Tang and Liu, 2009] is a social network among blog authors. Detailed information of these datasets is listed in Table 7.1.

7.3.2   BASELINE METHODS

We employ four NE models as baselines, including **DeepWalk, LINE, node2vec**, and **MNMF**. As mentioned in previous section, we implement CNRL on DeepWalk and node2vec. Taking DeepWalk for example; we denote the statistic-based and embedding-based community-enhanced DeepWalk as S-DW and E-DW, respectively. Similarly, the corresponding implementations of node2vec are denoted as S-n2v and E-n2v.

Besides, we also employ four popular link prediction methods as baselines, which are mainly based on local topological properties [Lü and Zhou, 2011]:

Common Neighbors (CN). For vertex v_i and v_j, CN [Newman, 2001] simply counts the common neighbors of v_i and v_j as similarity score: $sim(v_i, v_j) = |N_i^+ \cap N_j^+|$.

Salton Index. For vertex v_i and v_j, Salton index [Salton and McGill, 1986] further considers the degree of v_i and v_j. The similarity score can be formulated as: $sim(v_i, v_j) = (|N_i^+ \cap N_j^+|)/(\sqrt{|N_i^+| \times |N_j^+|})$.

Jaccard Index. For vertex v_i and v_j, Jaccard index is defined as: $sim(v_i, v_j) = (|N_i^+ \cap N_j^+|)/(|N_i^+ \cup N_j^+|)$.

Resource Allocation Index (RA). RA index [Zhou et al., 2009] is the sum of resources received by v_j: $sim(v_i, v_j) = \sum_{v_k \in N_i^+} \frac{1}{|N_k^+|}$.

For community detection, we select three most prominent methods as baselines.

Sequential Clique Percolation (SCP) [Kumpula et al., 2008] is a faster version of Clique Percolation [Palla et al., 2005], which detects communities by searching for adjacent cliques.

Link Clustering (LC) [Ahn et al., 2010] aims to find link communities rather than nodes.

BigCLAM [Yang and Leskovec, 2012] is a typical nonnegative matrix factorization-based model which can detect densely overlapping and hierarchically nested communities in massive networks.

7.3.3   PARAMETER SETTINGS AND EVALUATION METRICS

For a fair comparison, we apply the same representation dimension as 128 in all methods. In LINE, as suggested in Tang et al. [2015b], we set the number of negative samples to 5 and the learning rate to 0.025. We set the number of side samples to 1 billion for BlogCatalog and 10 million for other datasets. For random walk sequences, we set the walk length to 40, the window size to 5 and the number of sequences started from each vertex to 10. Besides, we employ grid-search to obtain the best performing hyper-parameters in MNMF.

Note that the representation vectors in CNRL consist of two parts, including the original vertex vectors and the corresponding community vectors. For fair comparison, we set the dimension of both vectors to 64 and finally obtain a 128-dimensional vector for each vertex. Besides, the smoothing factor α is set to 2 and β is set to 0.5.

For node classification, as each vertex in Cora, Citeseer, and Wiki has only one label, we employ L2-regularized logistic regression (L2R-LR), the default setting of Liblinear [Fan et al., 2008] to build classifiers. For multi-label classification in BlogCatalog, we train one-vs-rest logistic regression, and employ *micro-F1* for evaluation.

For link prediction, we employ a standard evaluation metric **AUC** [Hanley and McNeil, 1982]. Given the similarity of all vertex pairs, AUC is the probability that a random unobserved link has higher similarity than a random nonexistent link. Assume that we draw n independent comparisons, the AUC value is $(n_1 + 0.5n_2)/n$, where n_1 is the times that unobserved link has higher score and n_2 is the times that they have equal score.

Table 7.2: Node classification results (%)

Dataset	Cora		Citeseer		Wiki		BlogCatalog	
%Training ratio	10%	50%	10%	50%	10%	50%	1%	5%
DeepWalk	70.77	75.62	47.92	54.21	58.54	65.90	23.66	30.58
LINE	70.61	78.66	44.27	51.93	57.53	66.55	19.31	25.89
node2vec	73.29	78.40	49.47	55.87	58.93	66.03	24.47	30.87
MNMF	75.08	79.82	51.62	56.81	54.76	62.74	19.26	25.24
S-DW	74.14	80.33	49.72	57.05	59.72	67.75	23.80	30.25
E-DW	74.27	78.88	49.93	55.76	59.23	67.00	24.93	31.19
S-n2v	75.86	**82.81**	**53.12**	**60.31**	**60.66**	**68.92**	24.95	30.95
E-n2v	**76.30**	81.46	51.84	57.19	60.07	67.64	**25.75**	**31.13**

For community detection, we employ **modified modularity** [Zhang et al., 2015] to evaluate the quality of detected overlapping communities.

7.3.4 NODE CLASSIFICATION

In Table 7.2, we show the classification accuracies under different training ratios and different datasets. For each training ratio, we randomly select vertices as training set and the rest as test set. Note that, we employ smaller training ratios on BlogCatalog to accelerate the training of multi-label classifiers and evaluate the performance of CNRL under sparse scenes. From this table, we have the following observation.

(1) The proposed CNRL model consistently and significantly outperforms all baseline methods on node classification. Specifically, Community-enhanced DeepWalk outperforms DeepWalk, as well as Community-enhanced node2vec outperforms node2vec. It states the importance of incorporating community information and the flexibility of CNRL to various models. Moreover, with the consideration of community structure, CNRL is able to learn more meaningful and discriminative network representations and the learned representations are suitable to predictive tasks.

(2) While MNMF performs poorly on Wiki and BlogCatalog, CNRL performs reliably on all datasets. Furthermore, CNRL achieves more than 4% improvements than MNMF, although they both incorporate community information into NRL. It indicates that CNRL integrates community information more efficiently and has more stable performance than MNMF.

Table 7.3: Link prediction results (%)

Datasets	Cora	Citeseer	Wiki	BlogCatalog
CN	73.15	71.52	86.61	82.47
Salton	73.06	71.74	86.61	71.01
Jaccard	73.06	71.74	86.42	63.99
RA	73.25	71.74	86.98	**86.24**
DeepWalk	88.56	86.74	93.55	68.46
LINE	86.16	83.70	89.80	56.01
node2vec	92.99	89.57	93.49	67.31
MNMF	90.59	86.96	93.49	69.91
S-DW	89.67	87.39	93.43	68.62
E-DW	92.07	87.83	**94.99**	70.84
S-n2v	92.44	89.35	94.06	66.45
E-n2v	**93.36**	**89.78**	**94.99**	70.14

7.3.5 LINK PREDICTION

In Table 7.3, we show the AUC values of link prediction on different datasets while removing 5% edges. Note that, we show the results of LINE-1st on BlogCatalog, as it outperforms LINE. From this table, we observe the following.

(1) In most cases, NRL methods outperform traditional hand-crafted link prediction methods. It proves that NRL methods are effective to encode network structure into real-valued representation vectors. In this case, our proposed CNRL model consistently outperforms other NRL methods on different datasets. The results demonstrate the reasonability and effectiveness of considering community structure again.

(2) For BlogCatalog, the average degree (i.e., 32.39) of vertices is much larger than other networks, which will benefit simple statistic-based methods, such as CN and RA. Nevertheless, according to our experiments, when the network turns sparse by removing 80% edges, the performance of these simple methods will badly decrease (25% around). On the contrary, CNRL only decreases about 5%. It indicates that CNRL is more robust to the sparsity issue.

7.3.6 COMMUNITY DETECTION

We use modified modularity to evaluate the quality of detected communities. From Table 7.4, we observe that, S-CNRL (S-DW or S-n2v) is comparable with other state-of-the-art community detection methods, while E-CNRL (E-DW or E-n2v) significantly outperforms these baselines. It states that the communities detected by CNRL are meaningful under the mea-

Table 7.4: Community detection results

Datasets	SCP	LC	BigCLAM	S-DW	E-DW	S-n2v	E-n2v
Cora	0.076	0.334	0.464	0.464	**1.440**	0.447	1.108
Citeseer	0.055	0.315	0.403	0.486	**1.861**	0.485	1.515
Wiki	0.063	0.322	0.286	0.291	**0.564**	0.260	0.564

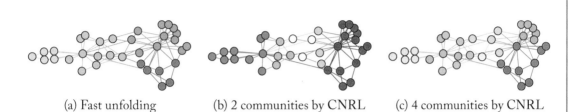

(a) Fast unfolding   (b) 2 communities by CNRL   (c) 4 communities by CNRL

Figure 7.2: Detected communities on Karate (Fast Unfolding, 2 communities by CNRL, 4 communities by CNRL).

surement of community quality. Moreover, it conforms to our assumptions about community assignment.

To summarize, all the results demonstrate the effectiveness and robustness of CNRL for incorporating community structure into node representations. It achieves consistent improvements comparing with other NRL methods on all network analysis tasks.

7.3.7 VISUALIZATIONS OF DETECTED COMMUNITIES

For a more intuitive sense of detected communities, we visualize the detected overlapping communities by CNRL on a toy network named Zachary's Karate network [Zachary, 1977] in Fig. 7.2. For comparison, we also show the detected non-overlapping communities by a typical community detection algorithm, Fast Unfolding [Blondel et al., 2008]. Note that we mark different communities with different colors, and use gradient color to represent the vertices belonging to multiple communities. From Fig. 7.2, we can see that CNRL is able to detect community structure with multiple scales, rather than clustering or partitioning vertices into fixed communities. Both the 2-community version and 4-community one are straightforward and reasonable according to the network structure.

7.4 FURTHER READING

Detecting communities from networks is a critical research filed in social science. In terms of community detection, traditional methods focus on partitioning the vertices into different groups, i.e., detecting non-overlapping communities. Existing non-overlapping community detection works mainly include clustering-based methods [Kernighan and Lin, 1970], modularity-based methods [Fortunato, 2010, Newman, 2006], spectral algorithms [Pothen et al., 1990], stochastic block models [Nowicki and Snijders, 2001], and so on. The major drawback of these traditional methods is that they cannot detect overlapping communities, which may not accord with real-world scenarios. To address this problem, CPM [Palla et al., 2005] is proposed to generate overlapping communities by merging overlapping k-cliques. Link clustering [Ahn et al., 2010] was proposed for overlapping community detection by employing non-overlapping community detection methods to partition the links instead of vertices and then assigning a single vertex to corresponding groups of its links.

During the last decade, community affiliation-based algorithms have shown their effectiveness on overlapping community detection [Wang et al., 2011, Yang and Leskovec, 2012, 2013]. Community affiliation-based algorithms predefine the number of communities and learn a vertex-community strength vector for each vertex and assign communities to vertices according to the vector. For example, Yang and Leskovec [2013] proposes Non-negative Matrix Factorization (NMF) method to approximate adjacency matrix A by FF^T, where matrix F is the vertex-community affinity matrix. Then the algorithm learns non-negative vertex embeddings and converts each dimension of the embeddings into a community. These community affiliation based algorithms try to approximate the adjacency matrix in value and design different objective functions for it. The idea of NMF was also introduced into NRL for community-structured graphs.

In terms of NRL, Wang et al. [2017g] proposes modularized nonnegative matrix factorization (MNMF) model to detect non-overlapping communities for improving vertex representations. However, there are two disadvantages of MNMF. First, it can only detect non-overlapping communities (i.e., each vertex only belongs to a specific community), which is usually not in conformity with real-world networks. Besides, MNMF is a matrix factorization-based model, and its optimization complexity is square to the number of vertices, which makes it hard to handle large-scale networks. In this chapter, we exploit the analogy between topics in text and communities in networks and propose a novel NRL model, CNRL, which can be easily and efficiently incorporated into existing NRL models. To the best of our knowledge, CNRL is the first attempt to learn community-enhanced network representations by utilizing the analogy between topics and communities.

There are also other extensions of NRL for community-structured graphs [Cavallari et al., 2017, Du et al., 2018, Lai et al., 2017]. For example, Cavallari et al. [2017] proposes ComE to learn vertex embedding and detect communities simultaneously. Specifically, each community in ComE is represented as a multivariate Gaussian distribution to model how its member vertices are distributed. With well-designed iterative optimization algorithms, ComE can be trained efficiently with the complexity that is linear to the graph size.

Part of this chapter was from our arxiv paper by Tu et al. [2016a].

CHAPTER 8

Network Embedding for Large-Scale Graphs

As there are many large-scale real-world networks, it's inefficient for existing NE approaches to store amounts of parameters in memory and update them edge after edge. With the knowledge that nodes having similar neighborhood will be close to each other in embedding space, we propose COSINE (Compressive NE) algorithm which reduces the memory footprint and accelerates the training process by parameters sharing among similar nodes. COSINE applies graph partitioning algorithms to networks and builds parameter sharing dependency of nodes based on the result of partitioning. With parameters sharing among similar nodes, COSINE injects prior knowledge about higher structural information into training process which makes NE more efficient and effective. COSINE can be applied to any *embedding lookup* method and learn high-quality embeddings with limited memory and shorter training time. We conduct experiments of multi-label classification and link prediction, where baselines and our model have the same memory usage. Compared with baseline methods, COSINE has up to 23% increase on classification and up to 25% increase on link prediction. Moreover, time of all representation learning methods using COSINE decreases from 30% to 70%.

8.1 OVERVIEW

As large-scale, online social networks such as Facebook, Twitter, and Sina Weibo are developing rapidly, a large-scale real-world network typically contains millions of nodes and billions of edges. Most existing NE algorithms do not scale for networks of this size due to three reasons. (1) The majority of NE algorithms rely on *embedding lookup* [Hamilton et al., 2017b] to build the embeddings for each node. We denote the set of nodes by V. The mapping function has a form as $f(v)=E \cdot v$, where v is the target node, $\mathbf{E} \in \mathbb{R}^{d \times |V|}$ is a matrix containing the embedding vectors for all nodes, d is the dimension of vectors, $|V|$ is the size of nodes and $\mathbf{v} \in \mathbb{I}_V$ is a one-hot indicator vector indicating the column of $\mathbf{E}$ corresponding to node v. When the size of nodes grows, the dimension of vectors needs to reduce to keep the memory usage not exceed the limit. On the assumption that we have a network containing 100 million nodes and each node is represented by a 128-dimension floating-point vector, the memory storage of $\mathbf{E}$ is more than 100 GB. As the dimension becomes fairly small, the parameters of a model cannot preserve enough information about original network and have poor performance on the downstream machine learning tasks. if a node has only a few edges to other nodes, chances are that the

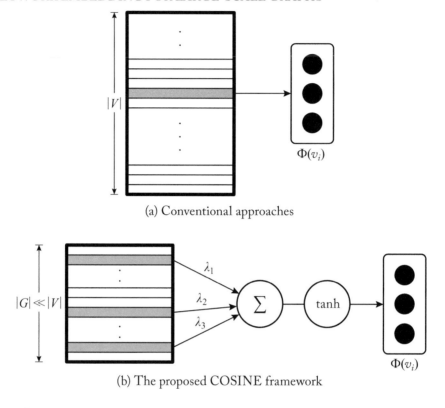

(a) Conventional approaches

(b) The proposed COSINE framework

Figure 8.1: Comparison between conventional approaches and COSINE on constructing embedding vectors, where $|\mathcal{G}| \ll |V|$.

training of the node's embedding would be insufficient. Broder et al. [2000] suggested that the distribution of degrees follows *power law*, which means there are many low-degree nodes in the large-scale network. (3) NE on large-scale networks needs to take a long time to train. However, many real-world networks are highly dynamic and evolving over time, so there is a need to speed up training process to follow that. To sum up, there is a challenge to improve the flexibility and efficiency of large-scale NE.

In this chapter, we explore how to share embedding parameters between nodes, which can address the computational inefficiency of *embedding lookup* methods and be a powerful form of regularization [Hamilton et al., 2017b]. We assume, in the network, there are groups containing nodes whose embeddings are partially close to each other. The groups of nodes can preserve the information of nodes effectively. Inspired by this, we propose a unified framework COSINE for compressive NE, which improves the quality of embeddings with limited memory and accelerates the training process of NE.

It is worth pointing out that COSINE can address all three problems of scalability above. First, parameters sharing can increase the dimension of vectors without extra memory usage. The dimension is very critical for preserving similarity from original networks. Second, one input training edge/pair can be used to update two nodes' parameters in previous methods while one edge/pair can be used to update several groups' parameters in COSINE, which affects more than two nodes. The low-degree nodes also have sufficient training, which solves the cold-start item problem. Third, the regularization of parameters sharing can be treated as prior knowledge about network structure, which reduces the number of training samples needed. Since there is a linear relation between running time and training samples, the training time will decrease with COSINE.

We apply COSINE to three state-of-the-art NE algorithms, DeepWalk [Perozzi et al., 2014], LINE[Tang et al., 2015b], and node2vec[Grover and Leskovec, 2016]. Then we conduct experiments on three large-scale real-world networks using the tasks of node classification and link prediction, where baselines and our model have the same memory usage. Experimental results show that COSINE significantly improves the performances of three methods, by up to 25% of AUC in link prediction and up to 23% of Micro-F1 in multi-label classification. Besides, COSINE greatly reduces the running time of these methods (30–70% decrease).

8.2   METHOD: COMPRESSIVE NETWORK EMBEDDING (COSINE)

In this section, we present a general framework which can cover several NE algorithms (including LINE, DeepWalk, and node2Vec), and learn better embeddings with limited memory resources. Our general framework consists of the following steps: (1) use Graph Partitioning methods to find vertices' partition/group from the network; (2) for each vertex, sample intelligently a set of groups to build the group mapping function; and (3) for each vertex, use an architecture based on GCN [Kipf and Welling, 2017] to aggregate information from its group set and output an embedding for each vertex; (4) use different NE objective functions to train the model. We will start by formalizing the problem of compressive NE and then explain each stage in detail. The framework is shown in Fig. 8.2.

8.2.1   PROBLEM DEFINITION

(Network Embedding) Given a network $G = (V, E)$, where V is the set of nodes and E is the set of edges, the goal of **network embedding** is to learn a mapping function $\Phi : V \mapsto \mathbb{R}^{|V| \times d}$, $d \ll |V|$. This mapping Φ defines the embedding of each node $v \in V$.

The parameter of the mapping function Φ in most existing approaches is a matrix which contains $|V| \times d$ elements, as they embed each node independently. However, there are many large networks which contain billions of nodes and their scales are still growing. It is difficult to store all embeddings in the memory when we train a large NE model. In this work, we

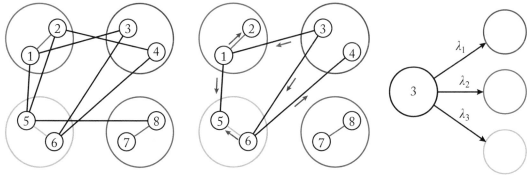

(a) Partitioning a simple graph   (b) Building group set by random walk   (c) Group set of node 3

Figure 8.2: A toy example of graph partitioning and group mapping. After partitioning, there are four groups in (a) which are presented in different colors. We begin several random walks rooted at Vertex 3 to sample its group set and set walk length as 2 in (b). According to the result of random walks, we find three groups which are related to Vertex 3 and use these three groups to present Vertex 3. Vertex 3 will share its embeddings parameters with vertices whose group sets also contain some of these three groups.

hypothesize that learning embeddings independently causes redundancy in the parameter, as the inter-similarity among nodes is ignored. For example, Jiawei Han and Philip S. Yu both belong to the Data Mining community in an academic network so that the embedding of Data Mining community can be the shared part of their embeddings. Therefore, we propose the problem of compressive NE to reduce the memory use by sharing parameters between similar nodes and make large NE available.

There are two main challenges of *compressive network embedding*: (1) how to find the shared and independent parts of embeddings between similar nodes; and (2) how to combine different parts of embeddings for each node to generate the node representation. For example, though Jiawei Han and Philip S. Yu may both have the embedding of Data Mining community in their shared part, the other portion in both vectors still has to be trained independently to capture extra information about their academic life. Following the intuition of creating partially shared embeddings, we represent each node v with a group set $S_v = (\mathcal{G}_1, \mathcal{G}_2, \mathcal{G}_3, \ldots, \mathcal{G}_M)$. Each group is a social unit where each node has a strong relation to each other and we denote the set of all groups by $\mathcal{G}$. We hypothesize that M groups are enough to figure out the characteristic of each node in networks. Having the same group set doesn't mean they have the same embedding, as different people have different preferences among groups. Therefore, the model also needs to learn the preference of each node while training. In summary, we formally define the *compressive network embedding* problem as follows:

(**Compressive Network Embedding**) Given a network $G = (V, E)$ and the dimension of embedding d, the goal of **compressive network embedding** is to learn a NE model which has less than $d|V|$ parameters (the traditional look-up methods need a memory space for $|V|$ d-dimension vectors). At the same time, the learned model can represent each node by a d-dimension vectors.

8.2.2   GRAPH PARTITIONING

Graph partitioning is used to divide a network into several partitions/groups, which is important for parameter sharing and incorporating high-order structure before training. There are explicit groups in social networks which consist of people who share similar characteristics and collectively have a sense of unity. As we don't have the information about existing social groups, we should use graph partitioning methods to assign each node one group. Based on the original groups, COSINE can sample more related group from network structure.

There are two kinds of methods which are able to assign groups to nodes: (1) overlapping methods like AGM [Yang and Leskovec, 2012], where a node can belong to multiple groups at once; (2) non-overlapping methods like Graph Coarsening and Graph Partitioning, where a node only belongs to one group. We want each node to have at most M different groups while overlapping methods can't limit the number of groups for each node. Thus, we choose non-overlapping methods in our framework.

HARP [Chen et al., 2017] and MILE [Liang et al., 2018] used Graph Coarsening to build a smaller network which approximates the global structure of its input. Learned embeddings from the small network serves as good initializations for representation learning in the input network. Graph Coarsening did not limit the number of origin nodes that belong to a coarsen group. It leads to unbalance among coarse groups. For instance, there may be some groups with only one node, which results in defective parameters sharing.

In our framework, we use Graph Partitioning method to assign each node a specific group. Graph Partitioning [Sanders and Schulz, 2011] is mostly used in *high performance computing* to partition the underlying graph model of computation and communication. Graph Partitioning divides nodes into several groups and encourages more edges in groups and fewer edges between groups. Each node links to the nodes in its group strongly and links to the rest nodes weakly. The advantage of graph partitioning is that it divides the nodes into k disjoint groups with roughly **equal sizes**, which benefits the parameter sharing.

8.2.3   GROUP MAPPING

After graph partitioning, we have a mapping $g(v)$ from one node to one group. In our framework, we plan to use a group set to represent a node instead of a single group. For each node, $g(v)$ plays an important role in the construction of group set S_v. However, we also need to find more related groups for each node.

We hypothesize that the neighbors' groups are characteristics for a node. There is proximity between nodes and their neighbors so that the neighbors' groups are also useful to present nodes. To introduce high order proximity into our model, we consider not only one-hop neighbors but also k-hop neighbors. We denote the j-th random walk rooted at vertex v_i as $W_{v_i}^j$. It is a stochastic process with random variables $v^1, v^2, \ldots, v^k$ such that $W_{v_i}^k$ is a k-hop neighbor for the root vertex v_i. There are two advantages of random walks for find neighbors' groups. First, random walks have been used to extract local structure of a network [Perozzi et al., 2014] and achieve great successes. Different from Breadth First Search (BFS), the walker can revisit a vertex in a walk which means this vertex is important for the local structure and we should pay more attention to it. Second, several walkers can walk simultaneously on a network. As we plan to solve the problem of compressive NE on large-scale networks, it is important for group mapping to run parallel.

We denote the number of groups in a node group set as $|S_v|$. After several walks rooted at v_i, we have a node set which contains the neighbors of v_i in k-hop. By the mapping $g(v)$, we have a group set S_{raw} which contains the neighbors' groups. In practice, the size of S_{raw} is usually bigger than $|S_v|$. Thus, we have to select the $|S_v|$ most related groups from S_{raw} according to the groups' frequency in walks. Algorithm 8.3 shows our group mapping algorithm. The function *Concatenate* is used to join two walk lists and the elements in W_{v_i} increased by walk length k. The function *SelectByFrequency* chooses the top n frequent items in the input list.

Figure 8.2 presents an illustrative example for graph partitioning and group mapping. Vertex 3 is far away from the pink group, as no walker can arrive at Vertex 7 and 8 in two hops. And there exist communications between the rest groups and Vertex 3 on the graph which is consistent with the mapping result. Therefore, group mapping can introduce high proximity of network structure to the mapping function Φ_v and motivate the partial embedding sharing. λ_i^v represents Vertex 3's preference for a specific group which is unknown to the mapping function and need to be learned.

8.2.4 GROUP AGGREGATION

Graph aggregation is the process where the model takes the group set as input and outputs the node embedding. We have a fixed number of groups for each node, which is convenient for us to aggregate group embeddings. However, we need a method to aggregate group embeddings into a single representation that remains invariant to the order of groups. In practice, there are various ways to aggregate embeddings that meet the requirement. Common methods include: (1) use an RNN to encode the group embedding one-by-one as a sequence, and augment with different permutations of the features so that ideally the RNN learns an order-invariant model for group aggregation; (2) use a symmetric function $f(g_1, g_2, \ldots, g_n)$, whose value does not change when the input $\{g_i | 1 \leq i \leq n\}$ is permuted. We now discuss these methods in detail.

Algorithm 8.3 Group Mapping(G, Φ_V)

Require: Graph $G(V, E)$ one-one mapping from nodes to groups $g(v)$ walk per vertex γ walk length k the number of groups in each set n

Ensure: mapping from nodes to group sets Φ_V

1: Initialize Φ_V
2: **for each** $v_i \in V$ **do**
3:     $W_{v_i} \leftarrow \{\}$
4:     **for** $j = 0$ to γ **do**
5:         $W_{v_i}^j \leftarrow RandomWalk(G, v_i, t)$
6:         $W_{v_i} \leftarrow Concatenate(W_{v_i}, W_{v_i}^j)$
7:     **end for**
8:     $\mathcal{G}_{v_i} \leftarrow g(W_{v_i})$
9:     $S_v \leftarrow SelectByFrequency(\mathcal{G}_{v_i}, n)$
10:     $\Phi_V(v_i) \leftarrow S_v$
11: **end for**

To train an order-invariant RNN model, we have to augment the training data by adding many random permutations. It leads to high computation overhead which is unacceptable for large networks.

Our framework aggregate group features via computing a weighted average of embeddings to estimate the optimal aggregation function. This method has been used in the GCN [Kipf and Welling, 2017] and GraphSAGE [Hamilton et al., 2017a]. For each node, it has an aggregation kernel whose size is equal to the number of groups $|S_v|$. We denote the kernel as $K_v = (\lambda_1^v, \lambda_2^v, \ldots, \lambda_{|S_v|}^v)$ where λ is a scalar learned from the network. To aggregate groups, we use the following equation:

$$f(g_1, g_2, \ldots, g_{|S_v|}) = \sum_{i=1}^{|S_v|} \lambda_i^v \Phi_{\mathcal{G}}(g_i), \tag{8.1}$$

where $\Phi_{\mathcal{G}}$ denotes the embedding mapping of groups, and there is no regularization on the sum of λ_i^v. To prevent the gradient explosion in the early stages of training, we use $\tanh$ as the activation function to regularize the entry value of $f(g_1, g_2, \ldots, g_{|S_v|})$. We have the final group aggregation function as follows:

$$f(S_v) = \tanh(\sum_{i=1}^{|S_v|} \lambda_i^v \Phi_{\mathcal{G}}(g_i)), \quad g_i \in S_v. \tag{8.2}$$

8.2.5   OBJECTIVE FUNCTION AND OPTIMIZATION

COSINE can apply to most existing NE algorithms by using the aggregated embeddings instead of the look-up embeddings. In this section, we take Skip-Gram objective function with Negative Sampling (SGNS) as an example to illustrate how COSINE tunes parameters of group embeddings and aggregation function via stochastic gradient descent. SGNS is the most common graph-based loss function:

$$\mathcal{L}(u, v) = -\log\left(\sigma\left(f_C(S_u)^\top f(S_v)\right)\right) - \sum_{i=1}^{K} E_{v_n \sim P_n(v)}\left[\log\left(\sigma\left(f_C(S_{v_n})^\top f(S_v)\right)\right)\right], \quad (8.3)$$

where u is a node that co-occurs near v (u and v co-occur at the same window in random-walk based methods and u is v's neighbor in LINE), σ is the sigmoid function, P_n is a negative sampling distribution, f_C is the aggregation function for context embedding, and K defines the number of negative samples. Importantly, unlike embedding look-up approaches, we not only share parameters among similar nodes via group embedding, but also use the same aggregation kernel while combining vertex embedding and context embedding [Mikolov et al., 2013b].

We adopt the asynchronous stochastic gradient algorithm (ASGD) [Recht et al., 2011] which is broadly used in learning embeddings for optimizing Eq. (8.3). In each step, we sample an edge (v_i, v_j) as a mini-batch and the gradient $w.r.t$ the group embedding e_g which belongs to one of the vertex v_j's groups will be calculated as:

$$\frac{\partial O}{\partial e_g} = -\left[1 - \sigma\left(f_C(S_{v_i})^\top f(S_{v_j})\right)\right] f_C(S_{v_i}) \otimes (1 - f(S_{v_j}) \otimes f(S_{v_j})) * \lambda_g^{v_j}, \quad (8.4)$$

where $\otimes$ is element-wise multiplication, and λ_g is the kernel parameter for group embedding e_g. To conduct a stable update on group embedding, we will adjust the learning rate according to the Euclidean norm of aggregation kernel. If the kernel norm is large, we decrease the learning rate to make the shared parameter change smoothly.

Besides computing the gradient of group embedding, we also need to compute the gradient of aggregation kernels for nodes:

$$\frac{\partial O}{\partial \lambda_g^{v_j}} = -\left[1 - \sigma\left(f_C(S_{v_i})^\top f(S_{v_j})\right)\right] \sum \left[f_C(S_{v_i}) \otimes (1 - f(S_{v_j}) \otimes f(S_{v_j})) \otimes e_g\right]. \quad (8.5)$$

We find if the group embedding e_g is similar to the node v_i's embedding, the aggregation kernel tend to put more weight on this group. After updating aggregation kernel, the model can learn the node's preference among groups. However, the gradient will explode when e_g and $f_C(S_{v_i})$ are nearly in the same direction. Instead of using gradient clipping, we consider the kernel gradient globally. If all gradients are large, we will decrease the learning rate, and vice versa.

Table 8.1: Networks used in our experiments

Name	YouTube	Flickr	Yelp
$|V|$	1,138,499	1,715,255	8,981,389
E	4,945,382	22,613,981	39,846,890
#Labels	47	20	22
Directed	directed	directed	undirected

8.3   EMPIRICAL ANALYSIS

We empirically evaluated the effectiveness and efficiency of COSINE. We applied the framework to three *embedding lookup* methods. According to the experimental results on three large-scale social network, our framework can improve the quality of embeddings with the same memory usage and reduce the running time.

To evaluate the quality of embeddings, we consider two network analysis tasks: multi-label classification and link prediction. We treat node embeddings as their features in the downstream machine learning tasks. The more beneficial the features are to the tasks, the better quality the embeddings have.

8.3.1   DATASETS

An overview of the networks we consider in our experiments is given in Table 8.1.

YouTube [Tang and Liu, 2009] contains 1,138,499 users and 4,945,382 social relations between them, which is crawled from the popular video sharing website. The labels represent groups of users that enjoy common video categories.

Flickr [Tang and Liu, 2009] contains 1,715,255 users and 22,613,981 social relations between them, which is crawled from the photo sharing website. The labels represent the interest groups of users such as "*black and white photos.*"

Yelp [Liang et al., 2018] contains 8,981,389 users and 39,846,890 social relations between them. The labels here represent the business categories on which the users have reviewed.

8.3.2   BASELINES AND EXPERIMENTAL SETTINGS

To demonstrate that COSINE can work with different graph embedding methods, we explore three popular state-of-the-art methods for NE.

LINE [Tang et al., 2015b] learns two separate network representations $LINE_{1st}$ and $LINE_{2nd}$, respectively. $LINE_{1st}$ can be only used on undirected networks and $LINE_{2nd}$ is suitable for undirected and directed networks. We choose LINE$_{2nd}$ as the baseline in our experiments.

DeepWalk [Perozzi et al., 2014] learns NEs from random walks. For each vertex, truncated random walks starting from the vertex are used to obtain the contextual information.

Node2vec [Grover and Leskovec, 2016] is an improved version of DeepWalk, where it generates random walks with more flexibility controlled through parameters p and q. We use the same setting as DeepWalk for those common hyper-parameters while employing a grid search over return parameter and in-out parameter $p, q \in \{0.25, 0.5, 1, 2, 4\}$.

Experimental Settings DeepWalk uses hierarchical sampling to approximate the softmax probabilities while hierarchical softmax is inefficient when compared with negative sampling [Mikolov et al., 2013b]. To embed large-scale network by DeepWalk, we switch to negative sampling, which is also used in node2vec and LINE, as Grover and Leskovec [2016] did. We set window size $w = 5$, random walk length $t = 40$ and walks per vertex $\gamma = 5$ for random-walk based methods. And, we use default settings [Tang et al., 2015b] for all hyper-parameters except the number of total training samples for LINE. We find these settings are effective and efficient for large-scale NE. With COSINE, network representation learning model can use less memory to learn embeddings with the same dimension. For instance, if we set the dimension of embeddings as d, we need $2d|V|$ floating-point numbers to store the uncompressed model and $|S_v||V| + 2d|\mathcal{G}| \approx |S_v||V|$ for the compressed one, where $|\mathcal{G}| \ll |V|$. And $|S_v|$ is also a small value, which means compressed model takes $\frac{|S_v|}{2d}$ times less space than uncompressed model. For a fair comparison, we use different dimensions in compressed and uncompressed models to make sure their memory usages are equal. We set $d = 100$ for uncompressed models and $d = 8$ for compressed models and adjust the number of groups for each dataset so that they have the same memory. Note that the size of group set for each node $|S_v|$ is 5 for all dataset, which is assumed to be enough for representing structural information.

We plan to prove that our framework can converge faster than the original ones. For *LINE*, we define one epoch means all edges have been trained for one time. For *DeepWalk* and *node2vec*, we regard feeding all walks for training as an iteration. We train each model with different epochs or iterations to find the best number of samples. For random walk based models with *COSINE* framework, we find the best number of samples is one iteration on part of the random walks. In other words, there is no need to do a complete iteration. For instance, 0.2 iteration means that the model just takes one iteration on 20% walks. As a framework, compressed and uncompressed model use the same random walks to ensure that the input data is the same.

8.3.3   LINK PREDICTION

Link prediction task can demonstrate the quality of NEs. Given a network, we randomly remove 10% links as the test set and the rest as the training set. We treat the training set as a new network

Table 8.2: AUC and MRR scores for link prediction

Metric	Algorithm	Dataset		
		YouTube	Flickr	Yelp
AUC	DeepWalk	0.926	0.927	0.943
	COSINE-DW	**0.941**	**0.968**	**0.951**
	node2vec	0.926	0.928	0.945
	COSINE-N2V	**0.942**	**0.971**	**0.953**
	LINE$_{2nd}$	0.921	0.934	0.943
	COSINE-LINE$_{2nd}$	**0.962**	**0.963**	**0.956**
MRR	DeepWalk	0.874	0.874	0.850
	COSINE-DW	**0.905**	**0.946**	**0.876**
	node2vec	0.874	0.876	0.857
	COSINE-N2V	**0.906**	**0.950**	**0.882**
	LINE$_{2nd}$	0.875	0.905	0.859
	COSINE-LINE$_{2nd}$	**0.939**	**0.935**	**0.892**

which is the input of network representation learning and employ the representations to compute the similarity scores between two nodes, which can be further applied to predict potential links between nodes.

We employ two standard link prediction metrics, AUC [Hanley and McNeil, 1982] and MRR [Voorhees et al., 1999], to evaluate compressed and uncompressed methods. We show the results of link prediction on different datasets in Table 8.2. The proposed COSINE framework consistently and significantly improves all baseline methods on link prediction, which means the high order proximity encoded before training is essential for measuring the similarity between nodes precisely.

8.3.4   MULTI-LABEL CLASSIFICATION

For multi-label classification task, we randomly select a portion of nodes as the training set and leave the rest as the test set. We treat NEs as node features and feed them into a one-vs-rest SVM classifier implemented by *LibLinear* [Fan et al., 2008]. We vary the training ratio from 1–10% to see the performance under sparse situation. To avoid overfitting, we train the classifier with L2-regularization.

We report the results under the best number of samples and compare the capacity in Tables 8.3, 8.4, and 8.5. We bold the results with higher performance between compressed and uncompressed models. From these tables, we have the following observations.

Table 8.3: Multi-label classification results in YouTube

	%Training ratio	1%	4%	7%	10%
Micro-F1(%)	DeepWalk	31.1%	35.9%	36.8%	37.4%
	COSINE-DW	**36.5%**	**42.0%**	**43.3%**	**44.0%**
	node2vec ($p = 2, q = 2$)	31.3%	36.5%	37.4%	38.0%
	COSINE-N2V ($p = 0.25, q = 0.5$)	**36.6%**	**41.8%**	**43.1%**	**44.1%**
	LINE$_{2nd}$	30.9%	34.7%	35.9%	36.2%
	COSINE-LINE$_{2nd}$	**36.3%**	**42.4%**	**43.6%**	**44.4%**
Macro-F1(%)	DeepWalk	14.0%	20.4%	22.5%	23.8%
	COSINE-DW	**21.2%**	**29.4%**	**31.7%**	**32.9%**
	node2vec ($p = 2, q = 2$)	14.3%	21.0%	22.9%	24.2%
	COSINE-N2V ($p = 0.25, q = 0.5$)	**21.2%**	**29.2%**	**31.7%**	**33.2%**
	LINE$_{2nd}$	14.1%	20.4%	22.5%	23.4%
	COSINE-LINE$_{2nd}$	**21.4%**	**31.3%**	**33.7%**	**35.0%**

(1) The proposed COSINE framework consistently and significantly improves all baseline methods on node classification. In YouTube, COSINE gives us at least 13% gain over all baselines in Micro-F1 and gives us at least 24% gain over all baselines in Macro-F1. In case of Flickr network, COSINE gives us at least 2% gain over all baselines in Micro-F1 and gives us at least 6% gain over all baselines in Macro-F1. As we can see in Yelp network, the classification scores don't change a lot with the growing training ratio, which can be explained by the weak relation between network structure information and node labels. The link prediction result has proven that COSINE can guarantee the high-quality of NEs in Yelp dataset.

(2) LINE$_{2nd}$ only considers the second order proximity in networks. As shown in previous work [Dalmia et al., 2018], LINE$_{2nd}$ has poor performance when the network is sparse like YouTube compared to DeepWalk and node2vec. COSINE encodes high-order proximity before training, which helps LINE$_{2nd}$ achieve the comparative performance in sparse networks.

(3) For node2vec with COSINE, the best return parameter q is not more than 1 in all networks, which means the local structure is less useful for the model training. Nodes in the same local structure share part of parameters while training. Therefore, there is no need to revisit neighbor nodes, which is consistent with our framework design.

Table 8.4: Multi-label classification results in Flickr

	%Training ratio	1%	4%	7%	10%
Micro-F1(%)	DeepWalk	39.7%	40.6%	40.9%	41.0%
	COSINE-DW	**40.4%**	**41.6%**	**42.1%**	**42.3%**
	node2vec ($p = 2, q = 0.5$)	39.8%	40.7%	40.9%	41.0%
	COSINE-N2V ($p = 1, q = 1$)	**40.4%**	**41.6%**	**42.1%**	**42.3%**
	LINE$_{2nd}$	**41.0%**	41.7%	41.8%	41.9%
	COSINE-LINE$_{2nd}$	40.8%	**42.1%**	**42.7%**	**42.9%**
Macro-F1(%)	DeepWalk	26.8%	29.9%	30.6%	31.0%
	COSINE-DW	**29.7%**	**33.6%**	**34.4%**	**34.9%**
	node2vec ($p = 2, q = 0.5$)	27.1%	30.1%	30.8%	31.2%
	COSINE-N2V ($p = 1, q = 1$)	**29.7%**	**33.6%**	**34.4%**	**34.9%**
	LINE$_{2nd}$	30.1%	32.8%	33.2%	33.4%
	COSINE-LINE$_{2nd}$	**32.0%**	**35.5%**	**36.2%**	**36.6%**

To summarize, COSINE framework effectively encodes high-order proximity before training, which is crucial for the parameter sharing. And the parameter sharing improve the capacity of baselines under the limited memory. Besides, COSINE is flexible to various social networks, whether they are sparse or dense. Moreover, it is a general framework, which works well with all baseline methods.

8.3.5 SCALABILITY

We now explore the scalability of our COSINE framework on three large-scale networks. As mentioned earlier, we should find the optimal sample number which makes the model to converge. A scalable model needs fewer samples to achieve better performance. In Fig. 8.3, we report the classification performance *w.r.t.* the training samples with 10% training ratio and the link prediction performance *w.r.t.* the training samples with dot product score function on YouTube network.

Figures 8.3a and 8.3b show that the classification performances of compressed and uncompressed models are the same without training data, which means graph partitioning results are useless at the beginning. Although the initial Macro-F$_1$ scores are the same, the score of

Table 8.5: Multi-label classification results in Yelp

	%Training ratio	1%	4%	7%	10%
Micro-F1(%)	DeepWalk	63.2%	63.3%	63.3%	63.3%
	COSINE-DW	**63.4%**	**63.8%**	**64.0%**	**64.0%**
	node2vec ($p = 0.5, q = 2$)	63.3%	63.4%	63.4%	63.4%
	COSINE-N2V ($p = 0.5, q = 2$)	**63.4%**	**63.8%**	**63.9%**	**64.0%**
	LINE$_{2nd}$	63.2%	63.3%	63.3%	63.3%
	COSINE-LINE$_{2nd}$	**63.4%**	**63.7%**	**63.8%**	**63.8%**
Macro-F1(%)	DeepWalk	34.6%	34.8%	34.8%	34.8%
	COSINE-DW	**36.0%**	**36.4%**	**36.5%**	**36.4%**
	node2vec ($p = 0.5, q = 2$)	35.0%	35.1%	35.1%	35.1%
	COSINE-N2V ($p = 0.5, q = 2$)	**36.1%**	**36.4%**	**36.5%**	**36.5%**
	LINE$_{2nd}$	35.1%	35.2%	35.3%	35.3%
	COSINE-LINE$_{2nd}$	**36.0%**	**36.2%**	**36.3%**	**36.2%**

compressed models grows faster with graph partitioning. Besides, compressed models converge faster than uncompressed models. Moreover, compressed models can converge and outperform uncompressed models using less training data, i.e., COSINE-LINE with 10 epochs gives us 40% gain over LINE with 90 epochs.

Figures 8.3c and 8.3d show that the link prediction performances of compressed and uncompressed models are different without training data, which means graph partitioning results are useful at the beginning. As the result of classification shows, compressed models also converge faster than uncompressed models while the growing speeds are nearly the same. We underline the following observations:

(1) Compressed models consistently and significantly reduce the training samples compared to the original models. It states the importance of incorporating graph partitioning results before training. Graph partitioning can encode high order proximity of the network, which is hard for baseline methods to learn.

(2) For models with COSINE framework, the best numbers of samples for two evaluation tasks are very close in a specific dataset while the best numbers are sometimes quite different for uncompressed models. It indicates that COSINE improves the stability of models among different tasks.

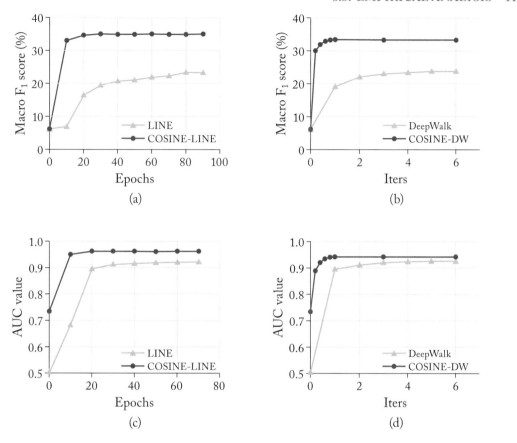

Figure 8.3: (a) and (b) show the classification performance *w.r.t.* the number of samples and (c) and (d) show the link prediction performance *w.r.t.* the number of samples. The dataset is YouTube.

Therefore, graph partitioning and parameters sharing help baselines to learn faster from data and reduce the need of training samples. We will discuss the time cost of graph partitioning and parameters sharing later in detail to show the time efficiency of COSINE.

8.3.6 TIME EFFICIENCY

In this section, we explore the time efficiency of our COSINE framework on three large-scale networks. We conduct NE on a modern machine with 12 cores. Figure 8.4a shows the running time to convergence of compressed and uncompressed models about LINE. We observe that COSINE significantly and consistently reduces the running time on three datasets by at least 20%. Figure 8.4b show the running time convergence of compressed and uncompressed models

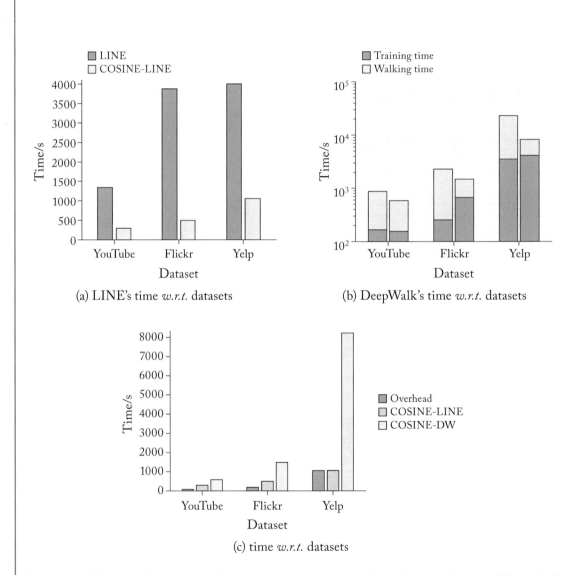

(a) LINE's time *w.r.t.* datasets

(b) DeepWalk's time *w.r.t.* datasets

(c) time *w.r.t.* datasets

Figure 8.4: Time performance on three large-scale networks. Note that in subfigure (b), the left part is DeepWalk and the right part is COSINE-DW.

for DeepWalk. There are training time and walking time for DeepWalk. As COSINE-DW need less walks for training to convergence, the walking times of COSINE-DW are also less than those of DeepWalk. COSINE also reduces the running time of DeepWalk, and the result of node2vec is similar to DeepWalk as they are both random-walk based methods. From Fig. 8.4, we observe that COSINE accelerates the training process of LINE and DeepWalk on three datasets by the reduction of training samples and the parameters sharing.

Besides the running time of compressed models, we examine the overhead of graph partitioning and group mapping. In Fig. 8.4, we report the time of overhead and two compressed models on three datasets. We observed that the influence of preprocessing is much smaller on YouTube and Flickr. In Yelp, the overhead is close to LINE's training time and is very small compared to COSINE-DW's time. When we add overhead to COSINE-LINE's time on Yelp, we find the total time is also reduced by 70%, which shows the time efficiency of COSINE. To sum up, the preprocessing, graph partitioning and group mapping, can significantly and consistently reduce the running time of baseline and the overhead of preprocessing has little influence on the total time.

8.3.7   DIFFERENT PARTITIONING ALGORITHMS

In this section, we examined three candidate partitioning algorithms and discuss the influence of different algorithms. They are:

KAFFPA [Sanders and Schulz, 2011] is a multi-level graph partitioning framework which contributes a number of improvements to the multilevel scheme which lead to enhanced partitioning quality. This includes flow-based methods, improved local search and repeated runs similar to the approaches used in multigrid solvers.

ParHIP [Meyerhenke et al., 2017] adapts the label propagation technique for graph clustering. By introducing size constraints, label propagation becomes applicable for both the coarsening and the refinement phase of multi-level graph partitioning.

mt-metis [LaSalle and Karypis, 2013] is the parallel version of METIS that not only improves the partitioning time but also uses much less memory.

We examine the performance of COSINE in YouTube dataset with different partitioning algorithms. We select $LINE_{2nd}$ as the base model and report the F_1 scores of classification, link prediction results in Fig. 8.5. We observed that the three algorithms have similar performance in classification tasks while ParHIP and mt-metis outperform KaFFPa in link prediction. In general, there are only minor differences between the performances of algorithms.

The mt-metis algorithm takes nearly 10% of KaFFPa's time to complete partitioning tasks. Thus, we select mt-metis as the partitioning algorithms in the COSINE framework for time efficiency. It takes less time and gives us a competitive performance.

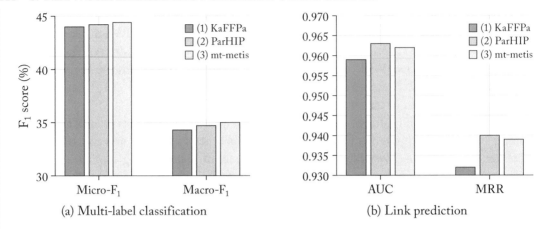

(a) Multi-label classification   (b) Link prediction

Figure 8.5: Performances of classification and link prediction on YouTube network *w.r.t.* graph partitioning algorithms.

8.4   FURTHER READING

In terms of NE methods that focus on efficiency, there are some work closely related to CO-SINE.

HARP [Chen et al., 2017] coarsens the graph and obtains a new graph made up of supernodes. Afterward, NE methods are applied to learn the representations of supernodes. The learned representations are used as the initialization of the supernodes' constituent nodes. Finally, the embedding methods will operate on more fine-grained subgraphs again for embedding learning. Compared with HARP, MILE [Liang et al., 2018], and GraphZoom [Deng et al., 2019] implement embedding refinement to learn better representations for nodes in lower computational cost and Akbas and Aktas [2019] directly use the supernode embeddings as node embeddings. RandNE [Zhang et al., 2018a] adopts Gaussian random projection technique to preserve the high-order proximity between nodes and reduce the time complexity. ProNE [Zhang et al., 2019a] builds a two-step paradigm for efficient NE. ProNE first initializes NEs by sparse matrix factorization, and then enhances embeddings by propagating them in the spectrally modulated space. NetSMF [Qiu et al., 2019] treats large-scale NE as sparse matrix factorization, which can be easily accelerated. While these methods still follow the setting of embedding lookup, our framework manages to reduce memory usage and improve the efficiency.

In general, the idea of reducing memory usage for efficiency is also related to model compression.

Model Compression focuses on building a light-weight approximation of the original model, whose size is reduced while preserving accuracy. Compression for CNN has been extensively studied and can be mainly divided into three following branches. (1) Low-rank matrix/tensor factorization [Denil et al., 2013, Jaderberg et al., 2014, Sainath et al., 2013] assumes a low-rank approximation for the weight matrices of both networks. (2) Network pruning [Han et al., 2015a,b, See et al., 2016, Zhang et al., 2017b] removes trivial weights in the neural network to make the network sparse. (3) Network quantization reduces the number of bits required to represent each weight, such as HashedNet [Chen et al., 2015] and QNN [Hubara et al., 2016]. There are also several techniques to compress word embeddings. Character-based neural language models [Botha et al., 2017, Kim et al., 2016] reduce the vocabulary size, but are hard to be adopted for Eastern Asian languages such as Chinese and Japanese which have a large vocabulary. To address this problem, Shu and Nakayama [2017] adopts various methods involving pruning and deep compositional coding to construct the embeddings with few basis vectors. Besides, Word2Bits [Lam, 2018] extends word2vec [Mikolov et al., 2013b] with a quantization function, showing that training with the function acts as a regularizer.

Part of this chapter was from our arxiv paper by Zhang et al. [2018b].

CHAPTER 9

Network Embedding for Heterogeneous Graphs

In reality, many networks are usually with multiple types of nodes and edges, widely known as heterogeneous information networks (HINs). HIN embedding aims to embed multiple types of nodes into a low-dimensional space. Although most HIN embedding methods consider heterogeneous relations in HINs, they usually employ one single model for all relations without distinction, which inevitably restricts the capability of NE. In this chapter, we take the structural characteristics of heterogeneous relations into consideration and propose a novel Relation structure-aware Heterogeneous Information Network Embedding model (RHINE). By exploring the real-world networks with thorough mathematical analysis, we present two structure-related measures which can consistently distinguish heterogeneous relations into two categories: Affiliation Relations (ARs) and Interaction Relations (IRs). To respect the distinctive characteristics of relations, we propose different models specifically tailored to handle ARs and IRs in RHINE, which can better capture the structures and semantics of the networks. Finally, we combine and optimize these models in a unified and elegant manner. Extensive experiments on three real-world datasets demonstrate that our model significantly outperforms previous methods in various tasks, including node clustering, link prediction, and node classification.

9.1   OVERVIEW

Networks in reality usually have multiple types of nodes and edges, and are recognized as HINs [Shi et al., 2017, Sun et al., 2011]. Taking the DBLP network for example, as shown in Fig. 9.1a, it contains four types of nodes: Author (A), Paper (P), Conference (C), and Term (T), and multiple types of relations: writing/written relations, and publish/published relations, etc. In addition, there are composite relations represented by meta-paths [Sun et al., 2011] such as *APA* (co-author relation) and *APC* (authors write papers published in conferences), which are widely used to exploit rich semantics in HINs. Thus, compared to homogeneous networks, HINs fuse more information and contain richer semantics. Directly applying traditional homogeneous models to embed HINs will inevitably lead to reduced performance in downstream tasks.

To model the heterogeneity of networks, several attempts have been done on HIN embedding. For example, some models employ meta-path based random walk to generate node sequences for optimizing the similarity between nodes [Dong et al., 2017, Fu et al., 2017, Shang

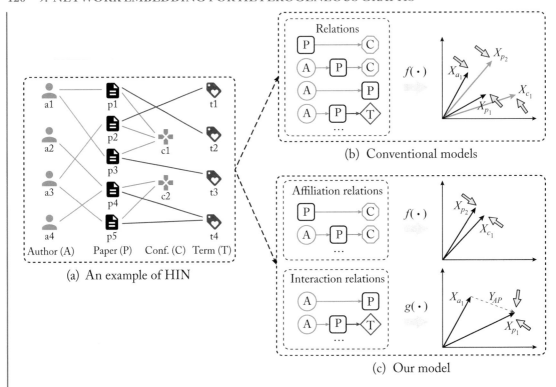

Figure 9.1: The illustration of an HIN and the comparison between conventional methods and our method (non-differentiated relations vs. differentiated relations).

et al., 2016]. Some methods decompose the HIN into simple networks and then optimize the proximity between nodes in each sub-network [Shi et al., 2018, Tang et al., 2015a, Xu et al., 2017]. There are also some neural network based methods that learn nonlinear mapping functions for HIN embedding [Chang et al., 2015, Han et al., 2018, Wang et al., 2018a, Zhang et al., 2017a]. Although these methods consider the heterogeneity of networks, they usually have an assumption that one single model can handle all relations and nodes, through keeping the representations of two nodes close to each other, as illustrated in Fig. 9.1b.

However, various relations in an HIN have significantly different structural characteristics, which should be handled with different models. Let's see a toy example in Fig. 9.1a. The relations in the network include atomic relations (e.g., *AP* and *PC*) and composite relations (e.g., *APA* and *APC*). Intuitively, *AP* relation and *PC* relation reveal rather different characteristics in structure. That is, some authors write some papers in the *AP* relation, which shows a peer-to-peer structure. While that many papers are published in one conference in the *PC* relation reveals the structure of one-centered-by-another. Similarly, *APA* and *APC* indicate peer-to-peer and one-centered-by-another structures, respectively. The intuitive examples clearly illustrate that relations in an HIN indeed have different structural characteristics.

It is non-trivial to consider different structural characteristics of relations for HIN embedding, due to the following challenges. (1) How to distinguish the structural characteristics of relations in an HIN? Various relations (atomic relations or meta-paths) with different structures are involved in an HIN. Quantitative and explainable criteria are desired to explore the structural characteristics of relations and distinguish them. (2) How to capture the distinctive structural characteristics of different categories of relations? Since the various relations have different structures, modeling them with one single model may lead to some loss of information. We need to specifically design appropriate models which are able to capture their distinctive characteristics. (3) The different models for the differentiated relations should be easily and smoothly combined to ensure simple optimization in a unified manner.

In this chapter, we present a novel model for HIN embedding, named **R**elation structure-aware **HIN** **E**mbedding (**RHINE**). In specific, we first explore the structural characteristics of relations in HINs with thorough mathematical analysis, and present two structure-related measures which can consistently distinguish the various relations into two categories: Affiliation Relations (ARs) with one-centered-by-another structures and Interaction Relations (IRs) with peer-to-peer structures. In order to capture the distinctive structural characteristics of the relations, we then propose two specifically designed models. For ARs where the nodes share similar properties [Yang and Leskovec, 2012], we calculate Euclidean distance as the proximity between nodes, so as to make the nodes directly close in the low-dimensional space. On the other hand, for IRs which bridge two compatible nodes, we model them as translations between the nodes. Since the two models are consistent in terms of mathematical form, they can be optimized in a unified and elegant way. We conduct comprehensive experiments to evaluate the performance of our model. Experimental results demonstrate that RHINE significantly outperforms previous NE models in various tasks.

9.2 METHOD: RELATION STRUCTURE-AWARE HIN EMBEDDING

In this section, we introduce some basic concepts and formalize the problem of HIN embedding. Then we present a novel RHINE, which individually handles two categories of relations (ARs and IRs) with different models in order to preserve their distinct structural characteristics, as illustrated in Fig. 9.1c.

9.2.1 PROBLEM FORMALIZATION

Heterogeneous Information Network (HIN). An HIN is defined as a graph $G = (V, E, T, \phi, \varphi)$, in which V and E are the sets of nodes and edges, respectively. Each node v and edge e are associated with their type mapping functions $\phi : V \to T_V$ and $\varphi : E \to T_E$, respectively. T_V and T_E denote the sets of node and edge types, where $|T_V| + |T_E| > 2$ and $T = T_V \cup T_E$.

Meta-path. A meta-path $m \in M$ is defined as a sequence of node types t_{v_i} or edge types t_{e_j} in the form of $t_{v_1} \xrightarrow{t_{e_1}} t_{v_2} \dots \xrightarrow{t_{e_l}} t_{v_{l+1}}$ (abbreviated as $t_{v_1} t_{v_2} \dots t_{v_{l+1}}$), which describes a composite relation between v_1 and v_{l+1}.

Node-Relation Triple. In an HIN G, relations R include atomic relations (e.g., links) and composite relations (e.g., meta-paths). A node-relation triple $\langle u, r, v \rangle \in P$, describes that two nodes u and v are connected by a relation $r \in R$. Here, P represents the set of all node-relation triples.

For example, as shown in Fig. 9.1a, $\langle a_2, APC, c_2 \rangle$ is a node-relation triple, meaning that a_1 writes a paper published in c_2.

Heterogeneous Information Network Embedding. Given an HIN $G = (V, E, T, \phi, \varphi)$, the goal of HIN embedding is to develop a mapping function $f : V \to \mathbb{R}^d$ that projects each node $v \in V$ to a low-dimensional vector in $\mathbb{R}^d$, where $d \ll |V|$.

9.2.2   DATA OBSERVATIONS

In this subsection, we first describe three real-world HINs and analyze the structural characteristics of relations in HINs. Then we present two structure-related measures which can consistently distinguish various relations quantitatively.

Dataset Description:

Before analyzing the structural characteristics of relations, we first briefly introduce three datasets used in this chapter, including DBLP,[1] Yelp,[2] and AMiner [Tang et al., 2008].[3] The detailed statistics of these datasets are illustrated in Table 9.1.

DBLP is an academic network, which contains four types of nodes: author (A), paper (P), conference (C), and term (T). We extract node-relation triples based on the set of relations {AP, PC, PT, APC, APT}. Yelp is a social network, which contains five types of nodes: user (U), business (B), reservation (R), service (S), and star level (L). We consider the relations {BR, BS, BL, UB, BUB}. AMiner is also an academic network, which contains four types of nodes, including author (A), paper (P), conference (C), and reference (R). We consider the relations {AP, PC, PR, APC, APR}. Notice that we can actually analyze all the relations based on meta-paths. However, not all meta-paths have a positive effect on embeddings [Sun et al., 2013]. Hence, following previous works [Dong et al., 2017, Shang et al., 2016], we choose the important and meaningful meta-paths.

---

[1] https://dblp.uni-trier.de
[2] https://www.yelp.com/dataset/
[3] https://www.aminer.cn/citation

Table 9.1: Statistics of the three datasets. t_u denotes the type of node u, $\langle u, r, v \rangle$ is a node-relation triple.

Datasets	Nodes	Number of nodes	Relations $(t_u \sim t_v)$	Number of relations	Avg. degree of t_u	Avg. degree of t_v	Measures D(r)	Measures S(r)	Relation category
DBLP	Term (T)	8,811	PC	14,376	1.0	718.8	718.8	0.05	AR
	Paper (P)	14,376	APC	24,495	2.9	2089.7	720.6	0.085	AR
	Author (A)	14,475	AP	41,794	2.8	2.9	1.0	0.0002	IR
	Conference (C)	20	PT	88,683	6.2	10.7	1.7	0.0007	IR
			APT	260,605	18.0	29.6	1.6	0.002	IR
Yelp	User (U)	1,286	BR	2,614	1.0	1307.0	1307.0	0.5	AR
	Service (S)	2	BS	2,614	1.0	1307.0	1307.0	0.5	AR
	Business (B)	2,614	BL	2,614	1.0	290.4	290.4	0.1	AR
	Star Level (L)	9	UB	30,838	23.9	11.8	2.0	0.009	IR
	Reservation (R)	2	BUB	528,332	405.3	405.3	1.0	0.07	IR
AMiner	Paper (P)	127,623	PC	127,623	1.0	1263.6	1264.6	0.01	AR
	Author (A)	164,472	APC	232,659	2.2	3515.6	1598.0	0.01	AR
	Reference (R)	147,251	AP	355,072	2.2	2.8	1.3	0.00002	IR
	Conference (C)	101	PR	392,519	3.1	2.7	1.1	0.00002	IR
			APR	1,084,287	7.1	7.9	1.1	0.00004	IR

Affiliation Relations and Interaction Relations:

In order to explore the structural characteristics of relations, we present mathematical analysis on the above datasets.

Since the degree of nodes can well reflect the structures of networks [Wasserman and Faust, 1994], we define a degree-based measure $D(r)$ to explore the distinction of various relations in an HIN. Specifically, we compare the average degrees of two types of nodes connected with the relation r, via dividing the larger one by the smaller one ($D(r) \geq 1$). Formally, given a relation r with nodes u and v (i.e., node relation triple $\langle u, r, v \rangle$), t_u and t_v are the node types of u and v, we define $D(r)$ as follows:

$$D(r) = \frac{\max [\bar{d}_{t_u}, \bar{d}_{t_v}]}{\min [\bar{d}_{t_u}, \bar{d}_{t_v}]}, \tag{9.1}$$

where $\bar{d}_{t_u}$ and $\bar{d}_{t_v}$ are the average degrees of nodes of the types t_u and t_v, respectively.

A large value of $D(r)$ indicates quite inequivalent structural roles of two types of nodes connected via the relation r (one-centered-by-another), while a small value of $D(r)$ means compatible structural roles (peer-to-peer). In other words, relations with a large value of $D(r)$ show much stronger affiliation relationships. Nodes connected via such relations share much more similar properties [Faust, 1997]. While relations with a small value of $D(r)$ implicate much stronger interaction relationships. Therefore, we call the two categories of relations as *Affiliation Relations* (ARs) and *Interaction Relations* (IRs), respectively.

In order to better understand the structural difference between various relations, we take the DBLP network as an example. As shown in Table 9.1, for the relation PC with $D(PC) = 718.8$, the average degree of nodes with type P is 1.0 while that of nodes with type C is 718.8. It shows that papers and conferences are structurally inequivalent. Papers are centered by conferences. While $D(AP) = 1.1$ indicates that authors and papers are compatible and peer-to-peer in structure. This is consistent with our common sense. Semantically, the relation PC means that "*papers are published in conferences*," indicating an affiliation relationship. Differently, AP means that "*authors write papers*," which explicitly describes an interaction relationship.

In fact, we can also define some other measures to capture the structural difference. For example, we compare the relations in terms of sparsity, which can be defined as:

$$S(r) = \frac{N_r}{N_{t_u} \times N_{t_v}}, \tag{9.2}$$

where N_r represents the number of relation instances following r. N_{t_u} and N_{t_v} mean the number of nodes with type t_u and t_v, respectively. The measure can also consistently distinguish the relations into two categories: ARs and IRs. The detailed statistics of all the relations in the three HINs are shown in Table 9.1.

Evidently, ARs and IRs exhibit rather distinct characteristics: (1) ARs indicate one-centered-by-another structures, where the average degrees of the types of end nodes are extremely different. They imply an affiliation relationship between nodes. (2) IRs describe peer-to-peer structures, where the average degrees of the types of end nodes are compatible. They suggest an interaction relationship between nodes.

9.2.3   BASIC IDEA

Through our exploration with thorough mathematical analysis, we find that the heterogeneous relations can be typically divided into ARs and IRs with different structural characteristics. In order to respect their distinct characteristics, we need to specifically design different while appropriate models for the different categories of relations.

For ARs, we propose to take Euclidean distance as a metric to measure the proximity of the connected nodes in the low-dimensional space. There are two motivations behind this: (1) First of all, ARs show affiliation structures between nodes, which indicate that nodes connected via such relations share similar properties [Faust, 1997, Yang and Leskovec, 2012]. Hence, nodes connected via ARs could be directly close to each other in the vector space, which is also consistent with the optimization of Euclidean distance [Danielsson, 1980]. (2) Additionally, one goal of HIN embedding is to preserve the high-order proximity. Euclidean distance can ensure that both first-order and second-order proximities are preserved as it meets the condition of the triangle inequality [Hsieh et al., 2017].

Different from ARs, IRs indicate strong interaction relationships between compatible nodes, which themselves contain important structural information of two nodes. Thus, we propose to explicitly model an IR as a translation between nodes in the low-dimensional vector space. Additionally, the translation based distance is consistent with the Euclidean distance in the mathematical form [Bordes et al., 2013]. Therefore, they can be smoothly combined in a unified and elegant manner.

9.2.4   DIFFERENT MODELS FOR ARS AND IRS

In this subsection, we introduce two different models exploited in RHINE for ARs and IRs, respectively.

Euclidean Distance for Affiliation Relations:

Nodes connected via ARs share similar properties [Faust, 1997], therefore nodes could be directly close to each other in the vector space. We take the Euclidean distance as the proximity measure of two nodes connected by an AR.

Formally, given an affiliation node-relation triple $\langle p, s, q \rangle \in P_{AR}$ where $s \in R_{AR}$ is the relation between p and q with weight w_{pq}, the distance between p and q in the latent vector space is calculated as follows:

$$f(p, q) = w_{pq}||\mathbf{X}_p - \mathbf{X}_q||_2^2, \tag{9.3}$$

in which $\mathbf{X}_p \in \mathbb{R}^d$ and $\mathbf{X}_q \in \mathbb{R}^d$ are the embedding vectors of p and q, respectively. As $f(p, q)$ quantifies the distance between p and q in the low-dimensional vector space, we aim to minimize $f(p, q)$ to ensure that nodes connected by an AR should be close to each other. Hence, we define the margin-based loss [Bordes et al., 2013] function as follows:

$$L_{EuAR} = \sum_{s \in R_{AR}} \sum_{\langle p,s,q \rangle \in P_{AR}} \sum_{\langle p',s,q' \rangle \in P'_{AR}} \max[0, \gamma + f(p, q) - f(p', q')], \tag{9.4}$$

where $\gamma > 0$ is a margin hyperparameter. P_{AR} is the set of positive affiliation node-relation triples, while P'_{AR} is the set of negative affiliation node-relation triples.

Translation-based Distance for Interaction Relations:

IRs demonstrate strong interactions between nodes with compatible structural roles. Thus, different from ARs, we explicitly model IRs as translations between nodes.

Formally, given an interaction node-relation triple $\langle u, r, v \rangle$ where $r \in R_{IR}$ with weight w_{uv}, we define the score function as:

$$g(u, v) = w_{uv}||\mathbf{X}_u + \mathbf{Y}_r - \mathbf{X}_v||, \tag{9.5}$$

where $\mathbf{X}_u$ and $\mathbf{X}_v$ are the node embeddings of u and v, respectively, and $\mathbf{Y}_r$ is the embedding of the relation r. Intuitively, this score function penalizes deviation of $(\mathbf{X}_u + \mathbf{Y}_r)$ from the vector $\mathbf{X}_v$.

For each interaction node-relation triple $\langle u, r, v \rangle \in P_{IR}$, we define the margin-based loss function as follows:

$$L_{TrIR} = \sum_{r \in R_{IR}} \sum_{\langle u,r,v \rangle \in P_{IR}} \sum_{\langle u',r,v' \rangle \in P'_{IR}} \max[0, \gamma + g(u, v) - g(u', v')], \tag{9.6}$$

where P_{IR} is the set of positive interaction node-relation triples, while P'_{IR} is the set of negative interaction node-relation triples.

9.2.5 A UNIFIED MODEL FOR HIN EMBEDDING

Finally, we smoothly combine the two models for different categories of relations by minimizing the following loss function:

$$
\begin{aligned}
L &= L_{EuAR} + L_{TrIR} \\
&= \sum_{s \in R_{AR}} \sum_{\langle p,s,q \rangle \in P_{AR}} \sum_{\langle p',s,q' \rangle \in P'_{AR}} \max[0, \gamma + f(p,q) - f(p',q')] \\
&\quad + \sum_{r \in R_{IR}} \sum_{\langle u,r,v \rangle \in P_{IR}} \sum_{\langle u',r,v' \rangle \in P'_{IR}} \max[0, \gamma + g(u,v) - g(u',v')].
\end{aligned}
$$

Sampling Strategy:

As shown in Table 9.1, the distributions of ARs and IRs are quite unbalanced. What's more, the proportion of relations are unbalanced within ARs and IRs. Traditional edge sampling may suffer from under-sampling for relations with a small amount or over-sampling for relations with a large amount. To address the problems, we draw positive samples according to their probability distributions. As for negative samples, we follow previous work [Bordes et al., 2013] to construct a set of negative node-relation triples $P'_{(u,r,v)} = \{(u',r,v)|u' \in V\} \cup \{(u,r,v')|v' \in V\}$ for the positive node-relation triple (u,r,v), where either the head or tail is replaced by a random node, but not both at the same time.

9.3 EMPIRICAL ANALYSIS

In this section, we conduct extensive experiments to demonstrate the effectiveness of RHINE.

9.3.1 DATASETS

As described in Section 9.2.2, we conduct experiments on three datasets, including DBLP, Yelp, and AMiner. The statistics of them are summarized in Table 9.1.

9.3.2 BASELINE METHODS

We compare our proposed model RHINE with six state-of-the-art NE methods.

- **DeepWalk** [Perozzi et al., 2014] performs a random walk on networks and then learns low-dimensional node vectors via the Skip-Gram model.

- **LINE** [Tang et al., 2015b] considers first-order and second-order proximities in networks. We denote the model that only uses first-order or second-order proximity as LINE-1st or LINE-2nd, respectively.

Table 9.2: Performance evaluation of node clustering

Methods	DBLP	Yelp	AMiner
DeepWalk	0.3884	0.3043	0.5427
LINE-1st	0.2775	0.3103	0.3736
LINE-2nd	0.4675	0.3593	0.3862
PTE	0.3101	0.3527	0.4089
ESim	0.3449	0.2214	0.3409
HIN2Vec	0.4256	0.3657	0.3948
metapath2vec	0.6065	0.3507	0.5586
RHINE	**0.7204**	**0.3882**	**0.6024**

- **PTE** [Tang et al., 2015a] decomposes an HIN to a set of bipartite networks and then learns the low-dimensional representation of the network.

- **ESim** [Shang et al., 2016] takes a given set of meta-paths as input to learn a low-dimensional vector space. For a fair comparison, we use the same meta-paths with equal weights in Esim and our model RHINE.

- **HIN2Vec** [Fu et al., 2017] learns the latent vectors of nodes and meta-paths in an HIN by conducting multiple prediction training tasks jointly.

- **Metapath2vec** [Dong et al., 2017] leverages meta-path based random walks and Skip-Gram model to perform node embedding. We leverage the meta-paths APCPA, UB-SBU, and APCPA in DBLP, Yelp, and AMiner, respectively, which perform best in the evaluations.

9.3.3   PARAMETER SETTINGS

For a fair comparison, we set the embedding dimension $d = 100$ and the size of negative samples $k = 3$ for all models. For DeepWalk, HIN2Vec, and metapath2vec, we set the number of walks per node $w = 10$, the walk length $l = 100$ and the window size $\tau = 5$. For RHINE, the margin γ is set to 1.

9.3.4   NODE CLUSTERING

We leverage K-means to cluster the nodes and evaluate the results in terms of normalized mutual information (NMI) [Shi et al., 2014]. As shown in Table 9.2, our model RHINE significantly outperforms all the compared methods. (1) Compared with the best competitors, the clustering performance of our model RHINE improves by 18.79%, 6.15%, and 7.84% on DBLP, Yelp, and

Table 9.3: Performance evaluation of link prediction

Methods	DBLP (A-A)		DBLP (A-C)		Yelp (U-B)		AMiner (A-A)		AMiner (A-C)	
	AUC	F1	AUC	F1	AUC	F1	AUC	F1	AUC	F1
DeepWalk	0.9131	0.8246	0.7634	0.7047	0.8476	0.6397	0.9122	0.8471	0.7701	0.7112
LINE-1st	0.8264	0.7233	0.5335	0.6436	0.5084	0.4379	0.6665	0.6274	0.7574	0.6983
LINE-2nd	0.7448	0.6741	0.8340	0.7396	0.7509	0.6809	0.5808	0.4682	0.7899	0.7177
PTE	0.8853	0.8331	0.8843	0.7720	0.8061	0.7043	0.8119	0.7319	0.8442	0.7587
ESim	0.9077	0.8129	0.7736	0.6795	0.6160	0.4051	0.8970	0.8245	0.8089	0.7392
HIN2Vec	0.9160	0.8475	0.8966	0.7892	0.8653	0.7709	0.9141	0.8566	0.8099	0.7282
metapath2vec	0.9153	0.8431	0.8987	0.8012	0.7818	0.5391	0.9111	0.8530	0.8902	0.8125
RHINE	**0.9315**	**0.8664**	**0.9148**	**0.8478**	**0.8762**	**0.7912**	**0.9316**	**0.8664**	**0.9173**	**0.8262**

AMiner, respectively. It demonstrates the effectiveness of our model RHINE by distinguishing the various relations with different structural characteristics in an HIN. In addition, it also validates that we utilize appropriate models for different categories of relations. (2) In all baseline methods, homogeneous NE models achieve the lowest performance, because they ignore the heterogeneity of relations and nodes. (3) RHINE significantly outperforms previous HIN embedding models (i.e., ESim, HIN2Vec, and metapath2vec) on all datasets. We believe the reason is that our proposed RHINE with appropriate models for different categories of relations can better capture the structural and semantic information of HINs.

9.3.5 LINK PREDICTION

We model the link prediction problem as a binary classification problem that aims to predict whether a link exists. In this task, we conduct co-author (*A-A*) and author-conference (*A-C*) link prediction for DBLP and AMiner. For Yelp, we predict user-business (*U-B*) links which indicate whether a user reviews a business. We first randomly separate the original network into training network and testing network, where the training network contains 80% relations to be predicted (i.e., *A-A*, *A-C*, and *U-B*) and the testing network contains the rest. Then, we train the embedding vectors on the training network and evaluate the prediction performance on the testing network.

The results of link prediction task are reported in Table 9.3 with respect to AUC and F1 score. It is clear that our model performs better than all baseline methods on three datasets. The reason behind the improvement is that our model based on Euclidean distance modeling relations can capture both the first-order and second-order proximities. In addition, our model RHINE distinguishes multiple types of relations into two categories in terms of their structural characteristics, and thus can learn better embeddings of nodes, which are beneficial for predicting complex relationships between two nodes.

Table 9.4: Performance evaluation of multi-class classification

Methods	DBLP		Yelp		AMiner	
	Macro-F1	Micro-F1	Macro-F1	Micro-F1	Macro-F1	Micro-F1
DeepWalk	0.7475	0.7500	0.6723	0.7012	0.9386	0.9512
LINE-1st	0.8091	0.8250	0.4872	0.6639	0.9494	0.9569
LINE-2nd	0.7559	0.7500	0.5304	0.7377	0.9468	0.9491
PTE	0.8852	0.8750	0.5389	0.7342	0.9791	0.9847
ESim	0.8867	0.8750	0.6836	0.7399	0.9910	0.9948
HIN2Vec	0.8631	0.8500	0.6075	0.7361	**0.9962**	**0.9965**
metapath2vec	0.8976	0.9000	0.5337	0.7208	0.9934	0.9936
RHINE	**0.9344**	**0.9250**	**0.7132**	**0.7572**	0.9884	0.9807

9.3.6   MULTI-CLASS CLASSIFICATION

In this task, we employ the same labeled data used in the node clustering task. After learning the node vectors, we train a logistic classifier with 80% of the labeled nodes and test with the remaining data. We use Micro-F1 and Macro-F1 score as the metrics for evaluation [Dong et al., 2017].

We summarize the results of classification in Table 9.4. As we can observe, (1) RHINE achieves better performance than all baseline methods on all datasets except Aminer. It improves the performance of node classification by about 4% on both DBLP and Yelp averagely. In terms of AMiner, the RHINE performs slightly worse than ESim, HIN2vec, and metapath2vec. This may be caused by over-capturing the information of relations *PR* and *APR* (*R* represents references). Since an author may write a paper referring to various fields, these relations may introduce some noise. (2) Although ESim and HIN2Vec can model multiple types of relations in HINs, they fail to perform well in most cases. Our model RHINE achieves good performance due to the respect of distinct characteristics of various relations.

9.3.7   COMPARISON OF VARIANT MODELS

In order to verify the effectiveness of distinguishing the structural characteristics of relations, we design three variant models based on RHINE as follows.

- **RHINE$_{Eu}$** leverages Euclidean distance to embed HINs without distinguishing the relations.

- **RHINE$_{Tr}$** models all nodes and relations in HINs with translation mechanism, which is just like TransE [Bordes et al., 2013].

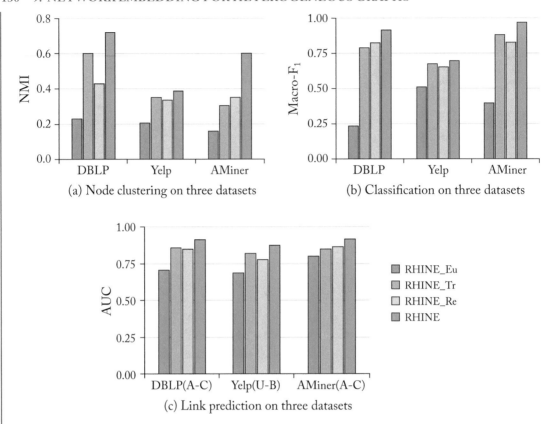

Figure 9.2: Performance evaluation of variant models.

- **RHINE$_{Re}$** leverages Euclidean distance to model IRs while translation mechanism for ARs, reversely.

We set the parameters of variant models as the same as those of our proposed model RHINE. The results of the three tasks are shown in Fig. 9.2. It is evident that our model outperforms RHINE$_{Eu}$ and RHINE$_{Tr}$, indicating that it is beneficial for learning the representations of nodes by distinguishing the heterogeneous relations. Besides, we find that RHINE$_{Tr}$ achieves better performance than RHINE$_{Eu}$. This is due to the fact that there are generally more peer-to-peer relationships (i.e., IRs) in the networks. Directly making all nodes close to each other leads to much loss of information. Compared with the reverse model RHINE$_{Re}$, RHINE also achieves better performance on all tasks, which implies that two models for ARs and IRs are well designed to capture their distinctive characteristics.

(a) DeepWalk   (b) metapath2vec   (c) RHINE

Figure 9.3: Visualization of node embeddings.

9.3.8   VISUALIZATION

To understand the representations of the networks intuitively, we visualize the vectors of nodes (i.e., papers) in DBLP learned with DeepWalk, metapath2vec, and RHINE in Fig. 9.3. As we can see, RHINE clearly clusters the paper nodes into four groups. It demonstrates that our model learns superior node embeddings by distinguishing the heterogeneous relations in HINs. In contrast, DeepWalk barely splits papers into different groups. Metapath2vec performs better than DeepWalk, but the boundary is blurry.

9.4   FURTHER READING

As a newly emerging network model, HINs can naturally model complex objects and their rich relations. HIN embedding, which aims to embed multiple types of nodes into a low-dimensional space, has received growing attention. Considerable researches have been done on representation learning for HINs [Chang et al., 2015, Chen et al., 2018, Dong et al., 2017, Fu et al., 2017, Han et al., 2018, Jacob et al., 2014, Shang et al., 2016, Shi et al., 2018, Tang et al., 2015a, Wang et al., 2018a, Xu et al., 2017, Zhang et al., 2017a]. Broadly speaking, these HIN embedding methods can be categorized into four types: random walk-based methods [Dong et al., 2017, Fu et al., 2017, Shang et al., 2016], decomposition-based methods [Shi et al., 2018, Tang et al., 2015a, Xu et al., 2017], deep neural network-based methods [Chang et al., 2015, Han et al., 2018, Wang et al., 2018a, Zhang et al., 2017a], and task-specific methods [Chen et al., 2018, Han et al., 2018, Jacob et al., 2014, Shi et al., 2018, Wang et al., 2018a].

Random Walk-Based Methods:
Random walk-based methods are inspired by word2vec [Mikolov et al., 2013a,b], where a node vector should be able to reconstruct the vectors of its neighborhood nodes which are defined by co-occurrence rate. Typically, metapath2vec [Dong et al., 2017] formalizes meta-path based random walks to construct heterogeneous neighborhoods of a node and leverages Skip-Gram of word2vec [Mikolov et al., 2013a] to learn the representation of networks. HIN2Vec [Fu et

al., 2017] conducts random walk and learns latent vectors of nodes and meta-paths by conducting multiple prediction training tasks jointly. Shang et al. propose ESim [Shang et al., 2016] to perform random walks based on user-defined meta-paths on HINs, and learn the vector representation of nodes appearing in the instance by maximizing the probability of meta-path instances.

Decomposition-Based Methods:

Decomposition-based methods separate an HIN into multiple simple homogeneous networks, and respectively embed these networks into low-dimensional spaces. As an extension of LINE, PTE [Tang et al., 2015a] is proposed to suit HIN embedding. It decomposes a HIN to a set of edgewise bipartite networks and then performs NE individually by using LINE. EOE [Xu et al., 2017] decomposes the complex academic heterogeneous network into a word co-occurrence network and an author cooperative network, and simultaneously performs representation learning on node pairs in sub-networks.

Deep Neural Network-Based Methods:

Deep neural network-based methods benefit from the powerful modeling capabilities of deep models, which employ different deep neural models, such as MLP, CNN, and Autoencoder, etc. to model heterogeneous data. For instance, HNE [Chang et al., 2015] utilizes CNN and MLP to extract the features of text and image data, respectively, and then projects different types of data into the same space through the transfer matrix to overcome the challenges of modeling heterogeneous data. SHINE [Wang et al., 2018a] uses the autoencoder to encode and decode the heterogeneous information in the social network, the emotional network and the portrait network, respectively, to obtain the feature representation, and then fuses these representations through an aggregate function to obtain the final node embeddings.

Task-specific Methods:

Task-specific methods mainly focus on solving a specific task (e.g., link prediction or recommendation) with representation learning on HINs. In order to predict links between nodes with different types in HINs, PME [Chen et al., 2018] projects different types of nodes into the same relation space and conducts heterogeneous link prediction. For recommendation in e-commerce, HERec [Shi et al., 2018] integrates matrix factorization with HIN embedding and predicts ratings for items. Fan et al. [2018] proposes a embedding model metagraph2vec, where both the structures and semantics are maximally preserved for malware detection.

To sum up, all the above-mentioned models deal with various relations without distinguishing their different properties and handle them with one single model. To the best of our knowledge, we make the first attempt to explore the different structural characteristics of relations in HINs and present two structure-related criteria which can consistently distinguish heterogeneous relations into ARs and IRs.

Part of this chapter was published in our AAAI19 conference paper by Lu et al. [2019].

PART IV

Network Embedding Applications

CHAPTER 10

Network Embedding for Social Relation Extraction

Conventional NE models learn low-dimensional node representations by simply regarding each edge as a binary or continuous value. However, there exists rich semantic information on edges and the interactions between vertices usually preserve distinct meanings, which are largely neglected by most existing models. In this chapter, NE plays as the main body of the proposed TransNet algorithm, which regards the interactions between vertices as a translation operation. Moreover, we formalize the task of Social Relation Extraction (SRE) to evaluate the capability of NE methods on modeling the relations between vertices. Experimental results on SRE demonstrate that TransNet significantly outperforms other baseline methods by 10% to 20% on *hits*@1.

10.1 OVERVIEW

The semantic information of edges in a graph was neglected in most existing NE models. As one of the essential network components, an edge is usually simplified as a binary or continuous value in conventional methods and most network analysis tasks. It is intuitive that such simplification cannot model rich information of edges well. It is also well-acknowledged that the interactions between vertices in real-world networks exhibit rich and variant meanings. For example, the following behaviors to the same user in social media may be caused by different reasons; two researchers co-author with a third one in an academic network due to various common interests. Therefore, it is essential to integrate the detailed relation information of edges into consideration, which is expected to enable extracting latent relations between vertices.

In this chapter, we propose the task of **S**ocial **R**elation **E**xtraction (SRE) to model and predict social relations for social networks. SRE is similar to the task of relation extraction in knowledge graphs (KG), for which the most widely-used methods are knowledge representation learning (KRL) such as TransE [Bordes et al., 2013]. The difference is that there are usually no well pre-defined relation categories in SRE, and relations between vertices are typically hidden in their interactive text (such as co-authored papers between two researchers). It is intuitive that social relations can be represented by key phrases extracted from the interactive text, and there are usually multiple relational labels to indicate the complex relation between two vertices.

SRE cannot be well addressed by existing NRL and KRL methods. Conventional NRL models ignore the rich semantic information on edges when learning node representations, while typical KRL models such as TransE only perform well when the relation between two entities is specifically annotated with a single label. According to our statistics, only 18% entity pairs in *FB15k* (a typical KG) possess multiple relation labels, while the percentages of multi-label edges in SRE datasets are severalfold. To address this issue, we present a novel translation-based NRL model **TransNet** to incorporate multiple relational labels on edges into NRL. "Translation" here means the movement that changes the position of a vector in representation space. Inspired by the successful utilization of translation analogy in word representation [Mikolov et al., 2013a] and KGs [Bordes et al., 2013], we embed vertices and edges into the same semantic space and employ a translation mechanism to deal with the interactions among them, i.e., the representation of tail vertex should be close to the representation of head vertex plus the representation of edge. To handle the multi-label scenario, in TransNet we design an auto-encoder to learn edge representations. Moreover, the decoder part can be utilized to predict labels of unlabeled edges.

We construct three network datasets for SRE, in which edges are annotated with a set of labels. Experimental results show that TransNet achieves significant and consistent improvements comparing with typical NRL models and TransE. It demonstrates that the proposed TransNet is efficient and powerful on modeling relationships between vertices and edges.

10.2   METHOD: TRANSNET

We start by formalizing the problem of SRE and then introduce the proposed model TransNet.

10.2.1   PROBLEM FORMALIZATION

SRE is similar to the task of relation extraction (RE) in KGs, which is an important technique that aims to extract relational facts to enrich existing KGs. Knowledge representation learning (KRL) methods such as TransE [Bordes et al., 2013], have been widely used for RE in KGs.

In this chapter, we present the task of SRE, which is designed to extract relations between social network vertices. Comparing with conventional RE in KGs, there are two main differences of SRE.

(1) In KGs, relation categories are usually well pre-defined, and relational facts are annotated precisely with human efforts. Conversely, SRE is proposed to deal with a new scenario, in which relations between social network vertices are latent and typically hidden in their interactive text information.

(2) In social networks, relations between vertices are extremely dynamic and complex, and cannot be portrayed well with a single label, because such practice cannot provide sufficient and accurate descriptions of these social relations. It is intuitive to represent the social relations by extracting key phrases from the interactive text information as a relation label set. These key phrases are flexible and capable of capturing the complex semantic information within social relations, as well as making these relations interpretable.

Formally, we define the problem of SRE as follows. Suppose there is a social network $G = (V, E)$, where V is the set of vertices, and $E \subseteq (V \times V)$ are edges between vertices. Besides, the edges in E are partially labeled, denoted as E_L. Without loss of generality, we define the relations between vertices as a set of labels, instead of a single label. Specifically, for each labeled edge $e \in E_L$, the label set of edge e is denoted as $l = \{t_1, t_2, \ldots\}$, where each label $t \in l$ comes from a fixed label vocabulary T.

Finally, given the overall network structure and the labeled edges in E_L, SRE aims to predict the labels over unlabeled edges in E_U, where $E_U = E - E_L$ represents the unlabeled edge set.

10.2.2 TRANSLATION MECHANISM

As shown in Fig. 10.1, TransNet consists of two critical components, i.e., translation part and edge representation construction part. In the following parts, we first give the detailed introduction of translation mechanism in TransNet. Afterward, we introduce how to construct the edge representations. At last, we give the overall objective function of TransNet.

Motivated by translation mechanisms in word representations [Mikolov et al., 2013a] and knowledge representations [Bordes et al., 2013], we assume that the interactions between vertices in social networks can also be portrayed as translations in the representation space.

Specifically, for each edge $e = (u, v)$ and its corresponding label set l, the representation of vertex v is expected to be close to the representation of vertex u plus the representation of edge e. As each vertex plays two roles in TransNet: head vertex and tail vertex, we introduce two vectors, $\mathbf{v}$ and $\mathbf{v}'$ for each vertex v, corresponding to its head representation and tail representation. After that, the translation mechanism among u, v and e can be formalized as

$$\mathbf{u} + \mathbf{l} \approx \mathbf{v}'. \tag{10.1}$$

Note that $\mathbf{l}$ is the edge representation obtained from l, which will be introduced in details in Section 10.2.3.

We employ a distance function $d(\mathbf{u} + \mathbf{l}, \mathbf{v}')$ to estimate the degree of (u, v, l) that matches Eq. (10.1). In practice, we simply adopt L_1-norm.

With the above definitions, for each (u, v, l) and its negative sample $(\hat{u}, \hat{v}, \hat{l})$, the translation part of TransNet aims to minimize the hinge-loss as follows:

$$\mathcal{L}_{trans} = \max(\gamma + d(\mathbf{u} + \mathbf{l}, \mathbf{v}') - d(\hat{\mathbf{u}} + \hat{\mathbf{l}}, \hat{\mathbf{v}}'), 0), \tag{10.2}$$

where $\gamma > 0$ is a margin hyper-parameter and $(\hat{u}, \hat{v}, \hat{l})$ is a negative sample from the negative sampling set N_e as

$$\begin{aligned} N_e = \{(\hat{u}, v, l) | (\hat{u}, v) \notin E\} &\cup \{(u, \hat{v}, l) | (u, \hat{v}) \notin E\} \\ &\cup \{(u, v, \hat{l}) | \hat{l} \cap l = \emptyset\}. \end{aligned} \tag{10.3}$$

In Eq. (10.3), the head vertex or tail vertex is randomly replaced by another disconnected vertex, and the label set is replaced by a non-overlapping label set.

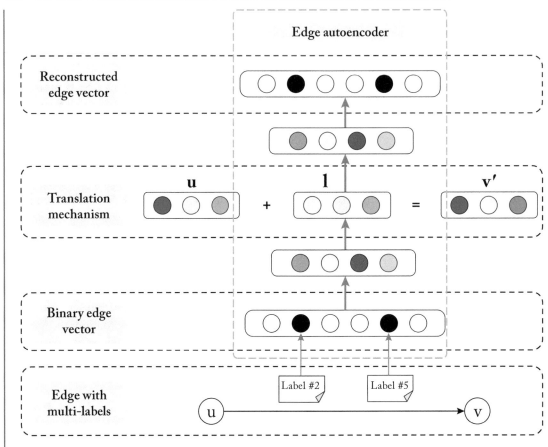

Figure 10.1: The framework of TransNet.

The vertex representations in Eq. (10.2) are treated as parameters, and the edge representations are generated from the corresponding label set, which will be introduced in the following part.

10.2.3   EDGE REPRESENTATION CONSTRUCTION

As shown in Fig. 10.1, we employ a deep autoencoder to construct the edge representations. The encoder part composes of several nonlinear transformation layers to transform the label set into a low-dimensional representation space. Moreover, the reconstruction process of the decoder part makes the representation preserve all the label information. In the following sections, we introduce how to realize it in detail.

Input Mapping: We first map the label set to an input vector of autoencoder. Specifically, for a label set $l = \{t_1, t_2, \ldots\}$ of edge e, we obtain a binary vector $\mathbf{s} = \{\mathbf{s}_i\}_{i=1}^{|T|}$, where $\mathbf{s}_i = 1$ if $t_i \in l$, and $\mathbf{s}_i = 0$ otherwise.

Nonlinear Transformation: Taking the obtained binary vector $\mathbf{s}$ as input, the encoder and decoder parts of the autoencoder consist of several nonlinear transformation layers as follows:

$$\begin{aligned} \mathbf{h}^{(1)} &= f(\mathbf{W}^{(1)}\mathbf{s} + \mathbf{b}^{(1)}), \\ \mathbf{h}^{(i)} &= f(\mathbf{W}^{(1)}\mathbf{h}^{(i-1)} + \mathbf{b}^{(i)}), i = 2, \ldots, K. \end{aligned} \tag{10.4}$$

Here, K represents the number of layers and f denotes the activation function. $\mathbf{h}^{(i)}$, $\mathbf{W}^{(1)}$ and $\mathbf{b}^{(i)}$ represent the hidden vector, transformation matrix and bias vector in the i-th layer, respectively. Specifically, we employ *tanh* activation function to get the edge representation $\mathbf{l} = \mathbf{h}^{(K/2)}$ as the vertex representations are real-valued, and *sigmoid* activation function to get the reconstructed output $\hat{\mathbf{s}}$ as the input vector $\mathbf{s}$ is binary.

Reconstruction Loss: Autoencoder aims to minimize the distance between inputs and the reconstructed outputs. The reconstruction loss is shown as:

$$\mathcal{L}_{rec} = ||\mathbf{s} - \hat{\mathbf{s}}||. \tag{10.5}$$

Here, we also adopt L_1-norm to measure the reconstruction distance, the same as in Eq. (10.2).

However, due to the sparsity of the input vector, the number of zero elements in $\mathbf{s}$ is much larger than that of non-zero elements. That means the autoencoder will tend to reconstruct the zero elements rather than non-zero ones, which is incompatible with our purpose. Therefore, we set different weights to different elements, and re-defined the loss function in Eq. (10.5) as follows:

$$\mathcal{L}_{ae} = ||(\mathbf{s} - \hat{\mathbf{s}}) \odot \mathbf{x}||, \tag{10.6}$$

where $\mathbf{x}$ is a weight vector and $\odot$ means the Hadamard product. For $\mathbf{x} = \{\mathbf{x}_i\}_{i=1}^{|T|}$, $\mathbf{x}_i = 1$ when $\mathbf{s}_i = 0$ and $\mathbf{x}_i = \beta > 1$ otherwise.

With the utilization of deep autoencoder, the edge representation not only remains the critical information of corresponding labels, but also has the ability of predicting the relation (labels) between two vertices.

10.2.4   OVERALL ARCHITECTURE

To preserve the translation mechanism among vertex and edge representations, as well as the reconstruction ability of edge representations, we combine the objectives in Eqs. (10.2) and (10.6), and propose a unified NRL model TransNet. For each (u, v, l) and its negative sample $(\hat{u}, \hat{v}, \hat{l})$, TransNet jointly optimizes the objective as follows:

$$\mathcal{L} = \mathcal{L}_{trans} + \alpha[\mathcal{L}_{ae}(l) + \mathcal{L}_{ae}(\hat{l})] + \eta\mathcal{L}_{reg}. \tag{10.7}$$

Here, we introduce two hyper-parameters α and η to balance the weights of different parts. Besides, $\mathcal{L}_{reg}$ is an $L2$-norm regularizer to prevent overfitting, which is defined as

$$\mathcal{L}_{reg} = \sum_{i=1}^{K}(||W^{(i)}||_2^2 + ||b^{(i)}||_2^2). \tag{10.8}$$

In order to prevent overfitting, we also employ dropout [Srivastava et al., 2014] to generate the edge representations. At last, we adopt Adam algorithm [Kingma and Ba, 2015] to minimize the objective in Eq. (10.7).

10.2.5   PREDICTION

With the learned vertex representations and the edge autoencoder, TransNet is capable of predicting the labels on the edges in E_U.

To be specific, given an unlabeled edge $(u, v) \in E_U$, TransNet assumes that the representations of u and v conform to the Eq. (10.1) with the potential edge representation. Thus, we can get the approximate edge representation through $\mathbf{l} = \mathbf{v}' - \mathbf{u}$. Naturally, we decode the edge representation $\mathbf{l}$ with the decoder part in Eq. (10.4) to obtain the predicted label vector $\hat{s}$. A larger weight $\hat{s}_i$ indicates t_i is more possible in l.

10.3   EMPIRICAL ANALYSIS

In order to investigate the effectiveness of TransNet on modeling relations between vertices, we compare the performance of our model to several baseline methods on SRE with three automatically constructed social network datasets.

10.3.1   DATASETS

ArnetMiner[1] [Tang et al., 2008] is an online academic website that provides search and mining services for researcher social networks. It releases a large-scale co-author network,[2] which consists of 1,712,433 authors, 2,092,356 papers, and 4,258,615 collaboration relations.

In this network, authors collaborate with different people on different topics, and the co-authored papers can reflect the detailed relations between them. Therefore, we construct the co-authored network with labeled edges in the following steps. First, we collect all the research interest phrases from the author profiles and build the label vocabulary with these phrases. These phrases are mainly crawled from the authors' personal home pages and annotated by themselves. Hence, these phrases are rather credible, which is also confirmed by our manual check. Second, for each co-author relationship, we filter out the in-vocabulary labels in the abstracts of co-authored papers and regard them as the ground truth labels of this edge. Note that, as the

[1]https://cn.aminer.org/
[2]https://cn.aminer.org/aminernetwork

Table 10.1: Datasets (ML indicates multi-label edges)

Datasets	Arnet-S	Arnet-M	Arnet-L
Vertices	187,939	268,037	945,589
Edges	1,619,278	2,747,386	5,056,050
Train	1,579,278	2,147,386	3,856,050
Test	20,000	300,000	600,000
Valid	20,000	300,000	600,000
Labels	100	500	500
ML proportion (%)	42.46	63.74	61.68

edges in co-author networks are undirected, we replace each edge with two directed edges with opposite directions.

Specifically, to better investigate the characteristics of different models, we construct three datasets with different scales, denoted as **Arnet-S**(small), **Arnet-M**(medium), and **Arnet-L**(large). The details are shown in Table 10.1.

10.3.2 BASELINES

We employ the following NRL models as baselines.

DeepWalk [Perozzi et al., 2014] performs random walks over networks to generate random walk sequences. With these sequences, it employs Skip-Gram [Mikolov et al., 2013a], an efficient word representation model, to learn vertex representations.

LINE [Tang et al., 2015b] defines the first-order and second-order proximities of networks and optimizes the joint and conditional probabilities of edges in large-scale networks.

node2vec [Grover and Leskovec, 2016] extends DeepWalk with a biased random walk strategy. It can explore the neighborhood architecture more efficiently.

For these NRL models, we treat SRE as a multi-label classification task. Therefore, we concatenate the head and tail vertex representations to form the feature vector and adopt one-vs-rest logistic regression implemented by Pedregosa et al. [2011] to train a multi-label classifier.

Besides, we also compare our model with the typical knowledge embedding model, **TransE** [Bordes et al., 2013]. For each training instance (u, v, l), where $l = \{t_1, t_2, \ldots\}$, it is intuitive to obtain several triples, i.e., (u, v, t_i) for each $t_i \in l$, which can be directly utilized to train TransE model. We adopt the similarity based predicting method as introduced in Bordes et al. [2013].

Table 10.2: SRE results on Arnet-S ($\times 100$ for $hits@k$, $\alpha = 0.5$ and $\beta = 20$)

Metric	hits@1	hits@5	hits@10	MeanRank	hits@1	hits@5	hits@10	MeanRank
DeepWalk	13.88	36.80	50.57	19.69	18.78	39.62	52.55	18.76
LINE	11.30	31.70	44.51	23.49	15.33	33.96	46.04	22.54
node2vec	13.63	36.60	50.27	19.87	18.38	39.41	52.22	18.92
TransE	39.16	78.48	88.54	5.39	57.48	84.06	90.60	4.44
TransNet	**47.67**	**86.54**	**92.27**	**5.04**	**77.22**	**90.46**	**93.41**	**4.09**

Table 10.3: SRE results on Arnet-M ($\times 100$ for $hits@k$, $\alpha = 0.5$ and $\beta = 50$)

Metric	hits@1	hits@5	hits@10	MeanRank	hits@1	hits@5	hits@10	MeanRank
DeepWalk	7.27	21.05	29.49	81.33	11.27	23.27	31.21	78.96
LINE	5.67	17.10	24.72	94.80	8.75	18.98	26.14	92.43
node2vec	7.29	21.12	29.63	80.80	11.34	23.44	31.29	78.43
TransE	19.14	49.16	62.45	25.52	31.55	55.87	66.83	23.15
TransNet	**27.90**	**66.30**	**76.37**	**25.18**	**58.99**	**74.64**	**79.84**	**22.81**

10.3.3   EVALUATION METRICS AND PARAMETER SETTINGS

For a fair comparison, we evaluate the performance for each triple (u, v, t_i) as in TransE, where $t_i \in l$. Besides, we also employ $hits@k$ and $MeanRank$ [Bordes et al., 2013] as evaluation metrics. Here, $MeanRank$ is the *mean* of predicted ranks of all annotated labels, while $hits@k$ means the proportion of correct labels ranked in the top k. Note that the above metrics will under-estimate the models that rank other correct labels in the same label set high. Hence, we can filter out these labels before ranking. We denote the primal evaluation setting as "Raw" and the latter one as "Filtered."

We set all the representation dimension to 100 for all models. In TransNet, we set the regularizer weight η to 0.001, the learning rate of Adam to 0.001 and the margin γ to 1. Besides, we employ a two-layer autoencoder for all datasets and select the best-performed hyper-parameters α and β on validation sets.

10.3.4   RESULTS AND ANALYSIS

Tables 10.2, 10.3, and 10.4 show the SRE evaluation results with different evaluation metrics on different datasets. In these tables, the left four metrics are raw results, and the right are filtered ones. From these tables, we have the following observations.

(1) Our proposed TransNet achieves consistent and significant improvements than all the baselines on all different datasets. More specifically, TransNet outperforms the best baseline,

Table 10.4: SRE results on Arnet-L ($\times 100$ for $hits@k$, $\alpha = 0.5$ and $\beta = 50$)

Metric	hits@1	hits@5	hits@10	MeanRank	hits@1	hits@5	hits@10	MeanRank
DeepWalk	5.41	16.17	23.33	102.83	7.59	17.71	24.58	100.82
LINE	4.28	13.44	19.85	114.95	6.00	14.60	20.86	112.93
node2vec	5.39	16.23	23.47	102.01	7.53	17.76	24.71	100.00
TransE	15.38	41.87	55.54	32.65	23.24	47.07	59.33	30.64
TransNet	**28.85**	**66.15**	**75.55**	**29.60**	**56.82**	**73.42**	**78.60**	**27.40**

i.e., TransE, by around 10–20% absolutely. It demonstrates the effectiveness and robustness of TransNet on modeling and predicting relations between vertices.

(2) All NRL models have poor performance on SRE task under various situations, due to the neglect of rich semantic information over edges when learning vertex representations. On the contrary, both TransE and TransNet incorporate this information into the learned representations, thus achieve promising results on SRE. It indicates the importance of considering the detailed edge information, as well as the rationality of translation mechanism on modeling relations between vertices.

(3) Comparing with TransNet, TransE also performs poorly as it can only consider a single label on an edge each time, which turns the representation of labels on the same edge to be identical. Such practice may accord with the scenario in knowledge graph, where only 18% entity pairs possess multiple relation labels, according to our statistics on *FB15k* [Bordes et al., 2013]. Conversely, the percentages of multi-label edges on SRE datasets are larger (42%, 64%, and 62% on Arnet-S, Arnet-M, and Arnet-L, respectively). Therefore, TransNet models all the labels over an edge simultaneously and can handle these issues well according to the results.

(4) TransNet has stable performance under different scales of networks. Moreover, when the number of labels turns larger, the performance of TransNet only has a small drop (from 90% to 80% on filtered $hits@10$), while NRL models and TransNet decrease more than 20%. This indicates the flexibility and stability of TransNet.

10.3.5   COMPARISON ON LABELS

To investigate the strengths of TransNet on modeling relations between vertices, we compare TransNet with TransE under high-frequency labels and low-frequency ones. In Table 10.5, we show the filtered $hits@k$ and *MeanRank* results on Arnet-S over top-5 labels and bottom-5 labels, respectively.

From this table, we find that TransE performs much better on high-frequency labels than low-frequency labels since there is a liberal quantity of training instances for high-frequency labels. Compared with TransE, TransNet has stable and worthy performance on both types of labels. The reason is that TransNet employs an autoencoder to construct edge representations,

Table 10.5: Label comparisons on Arnet-S ($\times 100$ for $hits@k$)

Tags	Top 5 labels				Bottom 5 labels			
Metric	*hits@1*	*hits@5*	*hits@10*	MeanRank	*hits@1*	*hits@5*	*hits@10*	MeanRank
TransE	58.82	85.68	91.61	**3.70**	52.21	82.03	87.75	5.65
TransNet	**77.26**	**90.35**	**93.53**	3.89	**78.27**	**90.44**	**93.30**	**4.18**

Table 10.6: Recommended top-3 labels for each neighbor

Neighbors	TransE	TransNet
Matthew Duggan	**ad hoc network;** wireless sensor network; wireless sensor networks	**management system;** **ad hoc network;** wireless sensor
K. Pelechrinis	**wireless network;** wireless networks; ad hoc network	**wireless network;** wireless sensor network; **routing protocol**
Oleg Korobkin	**wireless network;** wireless networks; **wireless communication**	resource management; **system design;** **wireless network**

which can take the correlations between labels into consideration. These correlations can provide additional information for low-frequency labels, which will benefit the modeling and predicting of them.

10.3.6   CASE STUDY

To demonstrate the effectiveness of TransNet, we provide a case in the test set of Arnet-S. The selected researcher is "A. Swami," and the recommended labels for its co-authors are shown in Table 10.6. In this table, labels in bold are correct ones. We observe that both TransE and TransNet can recommend reasonable labels to different neighbors, which can reflect their distinct co-author topics. However, for a specific neighbor, TransE only recommends similar labels due to its similarity-based recommendation method. Conversely, TransNet can recommend discriminative labels with the usage of decoder.

10.4   FURTHER READING

Only a small portion of NE methods are proposed to consider the rich semantics of edges and make detailed predictions of relations on edges. For example, SiNE [Wang et al., 2017f], SNE [Yuan et al., 2017], and SNEA [Wang et al., 2017e] learn node representations in signed networks, where each edge is either positive or negative. There is also a line of work [Chen et al., 2007, Ou et al., 2016, Zhou et al., 2017] focusing on learning NEs for directed graphs where each edge represents an asymmetric relationship. Nevertheless, such consideration of edges is rather simple and not suitable to other types of networks. To the best of our knowledge, we are the first to formalize the task of social relation extraction (SRE) to evaluate the capability of NRL methods on modeling the relations between vertices.

It's worth noting that relation extraction has been an essential task in knowledge graphs [Hoffmann et al., 2011, Lin et al., 2016, Mintz et al., 2009, Riedel et al., 2010, Surdeanu et al., 2012], which aims to extract relational facts to enrich existing KGs. This problem usually performs as relation classification, as there exist various large-scale KGs such as Freebase [Bollacker et al., 2008], DBpedia [Auer et al., 2007], and YAGO [Suchanek et al., 2007], with labeled relations between entities. However, there are usually no annotated explicit relations on edges in social networks, and it is also time-consuming to annotate edges in large-scale networks with human efforts. To address this issue, we propose to obtain the relations from interactive text information through natural language processing (NLP) techniques automatically.

How to model the relationships between vertices and edges is crucial for predicting the relations precisely. In word representation learning field, Mikolov et al. [2013a] found translation patterns such as "*King*" − "*Man*" = "*Queen*" − "*Woman*." In knowledge graphs, Bordes et al. [2013] interprets the relations as translating operations between head and tail entities in the representation space, i.e., "*head*" + "*relation*" = "*tail*." Note that knowledge graph embedding (KGE) [Wang et al., 2017d] is a hot research topic in NLP area, and the methods of KGE and NE are quite different.

Inspired by these analogies, we assume there also exists translation mechanism in social networks, and propose translation-based NRL model, TransNet. Extensions of TransNet includes the adaptation for semi-supervised settings [Tian et al., 2019], an efficient version for large-scale graphs [Yuan et al., 2018] and a joint learning of structural and semantic information of edges [Zheng et al., 2019].

Part of this chapter was published in our IJCAI17 conference paper by Tu et al. [2017b].

CHAPTER 11

Network Embedding for Recommendation Systems on LBSNs

The accelerated growth of mobile trajectories in location-based services brings valuable data resources to understand users' moving behaviors. Apart from recording the trajectory data, another major characteristic of these location-based services is that they also allow the users to connect whomever they like or are interested in. A combination of social networking and location-based services is called as location-based social networks (LBSN). Existing studies indicates a close association between social connections and trajectory behaviors of users in LBSNs. In order to better analyze and mine LBSN data, we jointly model social networks and mobile trajectories with the help of network embedding techniques, which serves as the key part of the entire algorithm. Our model consists of two components: the construction of social networks and the generation of mobile trajectories. First, we adopt a network embedding method for the construction of social networks. Second, we consider four factors that influence the generation process of mobile trajectories, namely user visit preference, influence of friends, short-term sequential contexts, and long-term sequential contexts. Finally, the two components are tied by sharing the user network embeddings. Experimental results on location and friend recommendation demonstrate the effectiveness of our model.

11.1   OVERVIEW

The accelerated growth of mobile usage brings a unique opportunity to data mining research communities. Among these rich mobile data, an important kind of data resource is the huge amount of mobile trajectory data obtained from GPS sensors on mobile devices. These sensor footprints provide a valuable information resource to discover users' trajectory patterns and understand their moving behaviors. Several location-based sharing services have emerged and received much attention, such as *Gowalla*[1] and *Brightkite*.[2]

Apart from recording user trajectory data, another major feature of these location-based services is that they also allow the users to connect whomever they like or are interested in. For

[1]https://en.wikipedia.org/wiki/Gowalla
[2]https://en.wikipedia.org/wiki/Brightkite

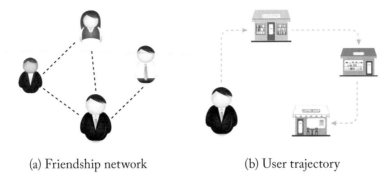

(a) Friendship network   (b) User trajectory

Figure 11.1: An illustrative example for the data in LBSNs. (a) Link connections represent the friendship between users. (b) A trajectory generated by a user is a sequence of chronologically ordered check-in records.

example, with *Brightkite* you can track on your friends or any other Brightkite users nearby using the phone's built in GPS. A combination of social networking and location-based services has lead to a specific style of social networks, termed as *location-based social networks (LBSN)* [Bao et al., 2012, Cho et al., 2011, Zheng, 2015]. We present an illustrative example for LBSNs in Fig. 11.1, and it can been seen that LBSNs usually include both the social network and mobile trajectory data.

Recent literature has shown that social link information is useful to improve existing recommendation tasks [Ma, 2014, Machanavajjhala et al., 2011, Yuan et al., 2014a]. Intuitively, users that often visit the same or similar locations are likely to be social friends and social friends are likely to visit same or similar locations. Specially, several studies have found that there exists close association between social connections and trajectory behaviors of users in LBSNs. On one hand, as shown in Cho et al. [2013], locations that are frequently visited by socially related persons tend to be correlated. On the other hand, trajectory similarity can be utilized to infer social strength between users [Pham et al., 2013, Zheng et al., 2011]. Therefore, we need to develop a comprehensive view to analyze and mine the information from the two aspects. In this chapter, our focus is to develop a joint approach to model LBSN data by characterizing both the social network and mobile trajectory data.

To conduct better and more effective data analysis and mining studies, there is a need to develop a joint model by capturing both network structure and trajectory behaviors on LBSNs. However, such a task is challenging. Social networks and mobile trajectories are heterogeneous data types. A social network is typically characterized by a graph, while a trajectory is usually modeled as a sequence of check-in records. A commonly used way to incorporate social connections into an application system (e.g., recommender systems) is to adopt the regularization techniques by assuming that the links convey the user similarity. In this way, the social connections are exploited as the side information but not characterized by a joint data model, and the model performance highly rely on the "homophily principle" of *like associates with like*.

In this chapter, we take the initiative to jointly model social networks and mobile trajectories using a neural network approach inspired by the recent progress on network embedding and RNN. Compared with other methods, neural network models can serve as an effective and general function approximation mechanism that is able to capture complicated data characteristics [Mittal, 2016]. In specific, our model consists of two components: the construction of social networks and the generation of mobile trajectories. We first adopt a network embedding method for the construction of social networks: a networking representation can be derived for a user. The key of our model lies in the component generating mobile trajectories. We have considered four factors that influence the generation process of mobile trajectories, namely user visit preference, influence of friends, short-term sequential contexts, and long-term sequential contexts. The first two factors are mainly related to the users themselves, while the last factors mainly reflect the sequential characteristics of historical trajectories. We set two different user representations to model the first two factors: a visit interest representation and a network representation. To characterize the last two contexts, we employ the RNN and GRU (a variant of RNN) models to capture the sequential relatedness in mobile trajectories at different levels, i.e., short term or long term. Finally, the two components are tied by sharing the user network representations: the information from the network structure is encoded in the user networking representation, which is subsequently utilized in the generation process of mobile trajectories.

To demonstrate the effectiveness of the proposed model, we evaluate our model using real-world datasets on two important LBSN applications, namely *next-location recommendation* and *friend recommendation*. For the first task, the trajectory data is the major information signal while network structure serves as auxiliary data. Our method consistently outperforms several competitive baselines. Interestingly, we have found that for users with little check-in data, the auxiliary data (i.e., network structure) becomes more important to consider. For the second task, the network data is the major information signal while trajectory data serves as auxiliary data. The finding is similar to that in the first task: our method still performs best, especially for those users with few friend links. Experimental results on the two important applications demonstrate the effectiveness of our model. In our approach, network structure and trajectory information complement each other. Hence, the improvement over baselines is more significant when either network structure or trajectory data is sparse.

11.2 METHOD: JOINT NETWORK AND TRAJECTORY MODEL (JNTM)

In this section, we will formalize the problems and present a novel neural network model for generating both social network and mobile trajectory data. In what follows, we study how to characterize each individual component. Then, we present the joint model followed by the parameter learning algorithm.

11.2.1   PROBLEM FORMALIZATION

We use L to denote the set of locations (a.k.a. *check-in* points or Point-Of-Interests (POIs)). When a user v checks in at a location l at the timestamp s, the information can be modeled as a triplet $\langle v, l, s \rangle$. Given a user v, her trajectory T_v is a sequence of triplets related to v: $\langle v, l_1, s_1 \rangle, ..., \langle v, l_i, s_i \rangle, ..., \langle v, l_N, s_N \rangle$, where N is the sequence length and the triplets are ordered by timestamps ascendingly. For brevity, we rewrite the above formulation of T_v as a sequence of locations $T_v = \{l_1^{(v)}, l_2^{(v)}, \ldots, l_N^{(v)}\}$ in chronological order. Furthermore, we can split a trajectory into multiple consecutive subtrajectories: the trajectory T_v is split into m_v subtrajectories $T_v^1, \ldots, T_v^{m_v}$. Each subtrajectory is essentially a subsequence of the original trajectory sequence. In order to split the trajectory, we compute the time interval between two check-in points in the original trajectory sequence, we follow Cheng et al. [2013] to make a splitting when the time interval is larger than six hours. To this end, each user corresponds to a trajectory sequence T_v consisting of several consecutive subtrajectories $T_v^1, \ldots, T_v^{m_v}$. Let T denote the set of trajectories for all the users.

Besides trajectory data, location-based services provide social connection links among users, too. Formally, we model the social network as a graph $G = (V, E)$, where each vertex $v \in V$ represents a user, each edge $e \in E$ represents the friendship between two users. In real applications, the edges can be either undirected or directed. As we will see, our model is flexible to deal with both types of social networks. Note that these links mainly reflect online friendship, which do not necessarily indicate that two users are friends in actual life.

Given the social network information $G = (V, E)$ and the mobile trajectory information T, we aim to develop a joint model which can characterize and utilize both kinds of data resources. Such a joint model should be more effective those built with a single data resource alone. In order to test the model performance, we set up two application tasks in LBSNs.

Task I. For the task of next-location recommendation, our goal is to recommend a ranked list of locations that a user v is likely to visit next at each step.

Task II. For the task of friend recommendation, our goal is to recommend a ranked list of users that are likely to be the friends of a user v.

We select these tasks because they are widely studied in LBSNs, respectively, representing two aspects for mobile trajectory mining and social networking analysis. Other tasks related to LBSN data can be equally solved by our model, which are not our focus in this chapter.

Before introducing the model details, we first summarize the used notations in this chapter in Table 11.1.

11.2.2   MODELING THE CONSTRUCTION OF THE SOCIAL NETWORK

In our task, we characterize the networking representations based on two considerations. First, a user is likely to have similar visit behaviors with their friends, and user links can be lever-

Table 11.1: Notations used in this paper

Notation	Descriptions
V, E	Vertex and edge set
L	Location set
T_v, T_v^j	Trajectory and the j-th subtrajectory of user v
m_v	Number of subtrajectories in the trajectory T_v of user v
$m_{v,j}$	Number of locations in the j-th subtrajectory of trajectory T_v of user v
$l_i^{(v,j)}$	The i-th location of the j-th subtrajectory of user v
U_{l_i}	Representation of location l_i used in representation modeling
U'_{l_i}	Representation of location l_i for prediction
P_v, F_v	Interest and friendship representation of user v
F'_v	Context friendship representation of user v
S_i	Short-term context representation after visiting location l_{i-1}
h_t	Long-term context representation after visiting location l_{t-1}

aged to share common visit patterns. Second, the networking structure is utilized as auxiliary information to enhance the trajectory modeling.

Formally, we use a d-dimensional embedding vector of use $F_v \in \mathbb{R}^d$ to denote the network representation of user v and matrix $F \in \mathbb{R}^{|V| \times d}$ to denote the network representations for all the users. The network representation is learned with the user links on the social network, and encodes the information for the structure patterns of a user.

The social network is constructed based on users' networking representations F. We first study how to model the generative probability for a edge of $v_i \to v_j$, formally as $\Pr[(v_i, v_j) \in E]$. The main intuition is that if two users v_i and v_j form a friendship link on the network, their networking representations should be similar. In other words, the inner product $F_{v_i}^\top \cdot F_{v_j}$ between the corresponding two networking representations will yield a large similarity value for two linked users. A potential problem will be such a formulation can only deal with undirected networks. In order to characterize both undirected and directed networks, we propose to incorporate a context representation for a user v_j, i.e., F'_{v_j}. Given a directed link $v_i \to v_j$, we model the representation similarity as $F_{v_i}^\top \cdot F'_{v_j}$ instead of $F_{v_i}^\top \cdot F_{v_j}$. The context representations are only used in the network construction. We define the probability of a link $v_i \to v_j$ by using a sigmoid function as follows:

$$\Pr[(v_i, v_j) \in E] = \sigma(-F_{v_i}^\top \cdot F_{v_j}') = \frac{1}{1 + \exp(-F_{v_i}^\top \cdot F_{v_j}')}. \tag{11.1}$$

When dealing with undirected networks, a friend pair (v_i, v_j) will be split into two directed links namely $v_i \rightarrow v_j$ and $v_j \rightarrow v_i$. For edges not existing in E, we propose to use the following formulation:

$$\Pr[(v_i, v_j) \notin E] = 1 - \sigma(-F_{v_i}^\top \cdot F_{v_j}') = \frac{\exp(-F_{v_i}^\top \cdot F_{v_j}')}{1 + \exp(-F_{v_i}^\top \cdot F_{v_j}')}. \tag{11.2}$$

Combining Eqs. (11.1) and (11.2), we essentially adopt a Bernouli distribution for modeling networking links. Following studies on networking representation learning [Perozzi et al., 2014], we assume that each user pair is independent in the generation process. That is to say the probabilities $\Pr[(v_i, v_j) \in E|F]$ are independent for different pairs of (v_i, v_j). With this assumption, we can factorize the generative probabilities by user pairs

$$\mathcal{L}(G) = \sum_{(v_i, v_j) \in E} \log \Pr[(v_i, v_j) \in E] + \sum_{(v_i, v_j) \notin E} \log \Pr[(v_i, v_j) \notin E]$$
$$= -\sum_{v_i, v_j} \log(1 + \exp(-F_{v_i}^\top \cdot F_{v_j}')) - \sum_{(v_i, v_j) \notin E} F_{v_i}^\top \cdot F_{v_j}'. \tag{11.3}$$

11.2.3   MODELING THE GENERATION OF THE MOBILE TRAJECTORIES

Note that a user trajectory is formatted as an ordered check-in sequences. Therefore, we model the trajectory generation process with a sequential neural network method. To generate a trajectory sequence, we generate the locations in it one by one conditioned on four important factors. We first summarize the four factors as below.

- *General visit preference*: A user's preference or habits directly determine her own visit behaviors.

- *Influence of Friends*: The visit behavior of a user is likely to be influenced by her friends. Previous studies [Cheng et al., 2012, Levandoski et al., 2012] indeed showed that socially correlated users tend to visit common locations.

- *Short-term sequential contexts*: The next location is closely related to the last few locations visited by a user. The idea is intuitive in that the visit behaviors of a user is usually related to a single activity or a series of related activities in a short time window, making that the visited locations have strong correlations.

- *Long-term sequential contexts*: It is likely that there exists long-term dependency for the visited locations by a user in a long time period. A specific case for long-term dependency will be periodical visit behaviors. For example, a user regularly has a travel for vocation in every summer vocation.

The first two factors are mainly related to the two-way interactions between users and locations. While the last two factors mainly reflect the sequential relatedness among the visited locations by a user.

Characterization of General Visit Preference:

We first characterize the general visit preference by the interest representations. We use a d-dimensional embedding vector of $P_v \in \mathbb{R}^d$ to denote the visit interest representation of user v and matrix $P \in \mathbb{R}^{|V| \times d}$ to denote the visit preference representations for all the users. The visit interest representation encodes the information for the general preference of a user over the set of locations in terms of visit behaviors.

We assume that one's general visit interests are relatively stable and does not vary too much in a given period. Such an assumption is reasonable in that a user typically has a fixed lifestyle (e.g., with a relatively fixed residence area) and her visiting behaviors are likely to show some overall patterns. The visit interest representation aims to capture and encode such visit patterns by using a d-dimensional embedding vector. For convenience, we call P_v as the *interest representation* for user v.

Characterization of Influence of Friends:

For characterizing influence of friends, a straightforward approach is to model the correlation between interest representations from two linked users with some regularization terms. However, such a method usually has high computational complexity. In this paper, we adopt a more flexible method: we incorporate the network representation in the trajectory generation process. Because the network representations are learned through the network links, the information from their friends are implicitly encoded and used. We still use the formulation of networking representation F_v introduced in Section 4.1.

Characterization of Short-Term Sequential Contexts:

Usually, the visited locations by a user in a short time window are closely correlated. A short sequence of the visited locations tend to be related to some activity. For example, a sequence "Home → Traffic → Office" refers to one's transportation activity from home to office. In addition, the geographical or traffic limits play an important role in trajectory generation process. For example, a user is more likely to visit a nearby location. Therefore, when a user decides what location to visit next, the last few locations visited by herself should be of importance for next-location prediction.

Based on the above considerations, we treat the last few visited locations in a short time window as the sequential history and predict the next location based on them. To capture the short-term visit dependency, we use the Recurrent Neural Network (RNN), a convenient way for modeling sequential data, to develop our model. Formally, given the j-th subsequence

$T_v^j = \{l_1^{(v,j)}, l_2^{(v,j)} \ldots l_{m_{v,j}}^{(v,j)}\}$ from the trajectory of user v, we recursively define the short-term sequential relatedness as follows:

$$S_i = \tanh(U_{l_{i-1}} + W \cdot S_{i-1}), \tag{11.4}$$

where $S_i \in \mathbb{R}^d$ is the embedding representation for the state after visiting location l_{i-1}, $U_{l_i} \in \mathbb{R}^d$ is the representation of location $l_i^{(v,j)}$ and $W \in \mathbb{R}^{d \times d}$ is a transition matrix. Here we call S_i *states* which are similar to those in Hidden Markov Models. RNN resembles Hidden Markov Models in that the sequential relatedness is also reflected through the transitions between two consecutive states. A major difference is that in RNN each hidden state is characterized by a d-dimensional embedding vector. The embedding vector corresponding to each state can be understood as an information summary until the corresponding location in the sequence.

Characterization of Long-Term Sequential Contexts:

In the above, short-term sequential contexts (five locations on average for our dataset) aim to capture the sequential relatedness in a short time window. The long-term sequential contexts are also important to consider when modeling trajectory sequences. For example, a user is likely to show some periodical or long-range visit patterns. To capture the long-term dependency, a straightforward approach will be to use another RNN model for the entire trajectory sequence. However, the entire trajectory sequence generated by a user in a long time period tends to contain a large number of locations, e.g., several hundred locations or more. A RNN model over long sequences usually suffers from the problem of "vanishing gradient."

To address the problem, we employ the Gated Recurrent Unit (GRU) for capturing long-term dependency in the trajectory sequence. Compared with traditional RNN, GRU incorporates several extra gates to control the input and output. Specifically, we use two gates in our model: input gate and forget gates. With the help of input and forget gates, the memory of GRU, i.e., the state C_t can remember the "important stuff" even when the sequence is very long and forget less important information if necessary. We present an illustrative figure for the architecture for recurrent neural networks with GRUs in Fig. 11.2.

Formally, consider the following location sequence $\{l_1, l_2, \ldots, l_m\}$, we denote the initial state by $C_0 \in \mathbb{R}^d$ and initial representation by $h_0 = \tanh(C_0) \in \mathbb{R}^d$. At a timestep of t, the new candidate state is updated as follows:

$$\widetilde{C_t} = \tanh(W_{c_1} U_{l_t} + W_{c_2} h_{t-1} + b_c), \tag{11.5}$$

where $W_{c_1} \in \mathbb{R}^{d \times d}$ and $W_{c_2} \in \mathbb{R}^{d \times d}$ are the model parameters, U_{l_t} is the embedding representation of location l_t which is the same representation used in short-term sequential relatedness, h_{t-1} is the embedding representation in the last step, and $b_c \in \mathbb{R}^d$ is the bias vector. Note that the computation of $\widetilde{C_t}$ remains the same as that in RNN.

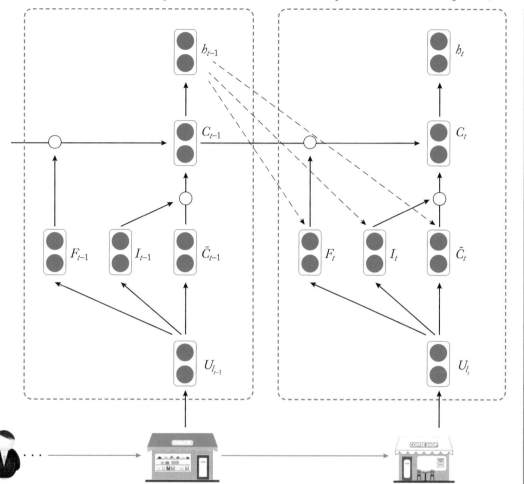

Figure 11.2: An illustrative architecture of recurrent neural networks with GRUs. Let $\widetilde{C_t}$ denote a candidate state. The current state C_t is a mixture of the last state C_{t-1} and the current candidate state $\widetilde{C_t}$. I_t and F_t are input and forget gate, respectively, which can control this mixture.

GRU does not directly replace the state with $\widetilde{C_t}$ as RNN does. Instead, GRU tries to find a balance between the last state C_{t-1} and a new candidate state $\widetilde{C_t}$:

$$C_t = i_t * \widetilde{C_t} + f_t * C_{t-1}, \tag{11.6}$$

where $*$ is entrywise product and i_t, $f_t \in \mathbb{R}^d$ are input and forget gate, respectively.

And the input and forget gates i_t, $f_t \in \mathbb{R}^d$ are defined as

$$i_t = \sigma(W_{i_1} U_{l_t} + W_{i_2} h_{t-1} + b_i) \tag{11.7}$$

and

$$f_t = \sigma(W_{f_1} U_{l_t} + W_{f_2} h_{t-1} + b_f), \tag{11.8}$$

where $\sigma(\cdot)$ is the sigmoid function, $W_{i_1}, W_{i_2} \in \mathbb{R}^{d \times d}$ and $W_{f_1}, W_{f_2} \in \mathbb{R}^{d \times d}$ are input and forget gate parameters, and $b_i, b_f \in \mathbb{R}^d$ are the bias vectors.

Finally, the representation of long-term interest variation at the timestep of t is derived as follows:

$$h_t = \tanh(C_t). \tag{11.9}$$

Similar to Eq. (11.4), h_t provides a summary which encodes the information till the t-th location in a trajectory sequence. We can recursively learn the representations after each visit of a location.

The Final Objective Function for Generating Trajectory Data:

Given the above discussions, we are now ready to present the objective function for generating trajectory data. Given the trajectory sequence $T_v = \{l_1^{(v)}, l_2^{(v)}, \ldots, l_m^{(v)}\}$ of user v, we factorize the log likelihood according to the chain rule as follows:

$$
\begin{aligned}
\mathcal{L}(T_v) &= \log \Pr[l_1^{(v)}, l_2^{(v)}, \ldots, l_m^{(v)} | v, \Phi] \\
&= \sum_{i=1}^{m} \log \Pr[l_i^{(v)} | l_1^{(v)}, \ldots, l_{i-1}^{(v)}, v, \Phi],
\end{aligned} \tag{11.10}
$$

where Φ denotes all the related parameters. As we can see, $\mathcal{L}(T_v)$ is characterized as a sum of log probabilities conditioned on the user v and related parameters Φ. Recall that the trajectory T_v is split into m_v subtrajectories $T_v^1, \ldots, T_v^{m_v}$. Let $l_i^{(v,j)}$ denote the i-th location in the j-th subtrajectory. The contextual locations for $l_i^{(v,j)}$ contain the preceding $(i-1)$ locations (i.e., $l_1^{(v,j)} \ldots l_{i-1}^{(v,j)}$) in the same subtrajectory, denoted by $l_1^{(v,j)} : l_{i-1}^{(v,j)}$, and all the locations in previous $(j-1)$ subtrajectories (i.e., $T_v^1, \ldots, T_v^{j-1}$), denoted by $T_v^1 : T_v^{j-1}$. With these notions, we can rewrite Eq. (11.10) as follows:

$$\mathcal{L}(T_v) = \sum_{i=1}^{m} \log \Pr[l_i^{(v,j)} | \underbrace{l_1^{(v,j)} : l_{i-1}^{(v,j)}}_{\text{short-term contexts}}, \underbrace{T_v^1 : T_v^{j-1}}_{\text{long-term contexts}}, v, \Phi]. \tag{11.11}$$

Given the target location $l_i^{(v,j)}$, the term of $l_1^{(v,j)} : l_{i-1}^{(v,j)}$ corresponds to the short-term contexts, the term of $T_v^1 : T_v^{j-1}$ corresponds to the long-term contexts, and v corresponds to the user context. The key problem becomes how to model the conditional probability $\Pr[l_i^{(v,j)} | l_1^{(v,j)} : l_{i-1}^{(v,j)}, T_v^1 : T_v^{j-1}, v, \Phi]$.

For short-term contexts, we adopt the RNN model described in Eq. (11.4) to characterize the location sequence of $l_1^{(v,j)} : l_{i-1}^{(v,j)}$. We use S_i^j to denote the derived short-term representation after visiting the i-th location in the j-th subtrajectory; for long-term contexts, the

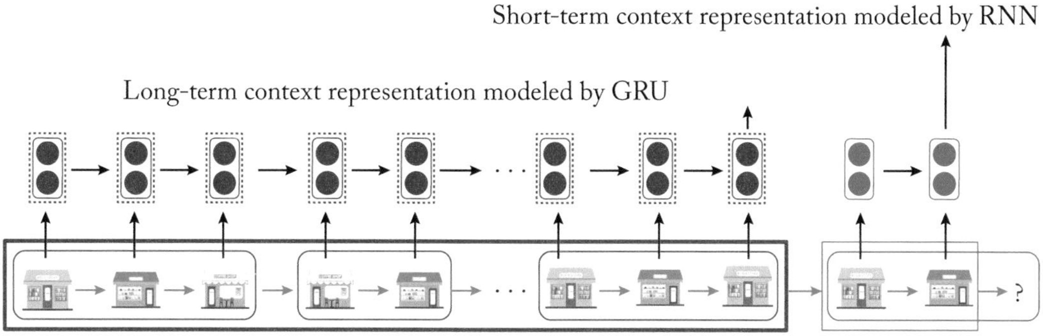

Figure 11.3: An illustrative figure for modeling both short-term and long-term sequential contexts. The locations in a rounded rectangular indicates a subtrajectory. The locations in red and blue rectangular are used for long-term and short-term sequential contexts, respectively. "?" is the next location for prediction.

locations in the preceding subtrajectories $T_v^1 \ldots T_v^{j-1}$ are characterized using the GRU model in Eqs. (11.6)–(11.9). We use h^j to denote the derived long-term representation after visiting the locations in first j subtrajectories. We present an illustrative example for the combination of short-term and long-term contexts in Fig. 11.3.

So far, given a target location $l_i^{(v,j)}$, we have obtained four representations corresponding to the four factors: networking representation (i.e., F_v), visit interest representation (i.e., P_v), short-term context representation S_{i-1}^j, and long-term context representation h^{j-1}. We concatenate them into a single context representation $R_v^{(i,j)} = [F_v; P_v; S_{i-1}^j; h^{j-1}] \in \mathbb{R}^{4d}$ and use it for next-location generation. Given the context representation $R_v^{(i,j)}$, we define the probability of $l_i^{(v,j)}$ as

$$
\begin{aligned}
&\Pr[l_i^{(v,j)}|l_1^{(v,j)} : l_{i-1}^{(v,j)}, T_v^1 : T_v^{j-1}, v, \Phi] \\
=\ &\Pr[l_i^{(v,j)}|R_v^{(i,j)}] \\
=\ &\frac{\exp(R_v^{(i,j)} \cdot U'_{l_i^{(v,j)}})}{\sum_{l \in L} \exp(R_v^{(i,j)} \cdot U'_l)},
\end{aligned}
\tag{11.12}
$$

where parameter $U'_l \in \mathbb{R}^{4d}$ is location representation of location $l \in L$ used for prediction. Note that this location representation U'_l is totally different with the location representation $U_l \in \mathbb{R}^d$ used in short-term and long-term context modeling. The overall log likelihood of trajectory generation can be computed by adding up all the locations.

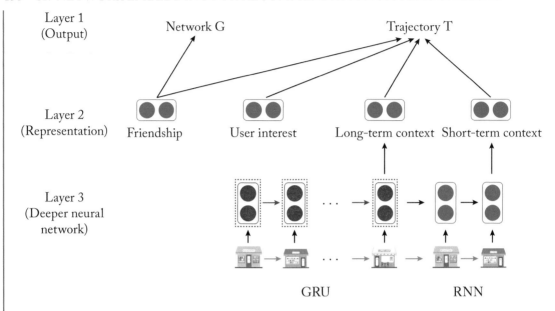

Figure 11.4: An illustrative architecture of the proposed model JNTM.

11.2.4 THE JOINT MODEL

Our general model is a linear combination between the objective functions for the two parts. Given the friendship network of $G = (V, E)$ and user trajectory T, we have the following log likelihood function:

$$\begin{aligned} \mathcal{L}(G, T) &= \mathcal{L}_{\text{network}}(G) + \mathcal{L}_{\text{trajectory}}(T) \\ &= \mathcal{L}(G) + \sum_{v \in V} \mathcal{L}(T_v), \end{aligned} \qquad (11.13)$$

where $\mathcal{L}_{\text{network}}(G)$ is defined in Eq. (11.3) and $\mathcal{L}_{\text{trajectory}}(T) = \sum_{v \in V} \mathcal{L}(T_v)$ is defined in Eq. (11.11), respectively. We name our model as *Joint Network and Trajectory Model (JNTM)*.

We present an illustrative architecture of the proposed model JNTM in Fig 11.4. Our model is a three-layer neural network for generating both social network and user trajectory. In training, we require that both the social network and user trajectory should be provided as the objective output to the train the model. Based on such data signals, our model naturally consists of two objective functions. For generating the social network, a network-based user representation was incorporated; for generating the user trajectory, four factors were considered: network-based representation, general visiting preference, short-term and long-term sequential contexts. These two parts were tied by sharing the network-based user representation.

11.2.5 PARAMETER LEARNING

Now we will show how to train our model and learn the parameters, i.e., user interest representation $P \in \mathbb{R}^{|V| \times d}$, user friendship representation $F, F' \in \mathbb{R}^{|V| \times d}$, location representations $U \in \mathbb{R}^{|L| \times d}, U' \in \mathbb{R}^{|L| \times 4d}$, initial short-term representation $S_0 \in \mathbb{R}^d$, transition matrix $W \in \mathbb{R}^{d \times d}$, initial GRU state $C_0 \in \mathbb{R}^d$ and GRU parameters $W_{i_1}, W_{i_2}, W_{f_1}, W_{f_2}, W_{c_1}, W_{c_2} \in \mathbb{R}^{d \times d}, b_i, b_f, b_c \in \mathbb{R}^d$.

Negative Sampling. Recall that the log likelihood of network generation Eq. (11.3) includes $|V| \times |V|$ terms. Thus, it takes at least $O(|V|^2)$ time to compute, which is time-consuming. Therefore, we employ negative sampling technique which is commonly used in NLP area [Mikolov et al., 2013b] to accelerate our training process.

Note that real-world networks are usually sparse, i.e., $O(E) = O(V)$. The number of connected vertex pairs (positive examples) are much less than the number of unconnected vertex pairs (negative examples). The core idea of negative sampling is that most vertex pairs serve as negative examples and thus we don't need to compute all of them. Instead, we compute all connected vertex pairs and n_1 random unconnected vertex pairs as an approximation where $n_1 \ll |V|^2$ is the number of negative samples. In our settings, we set $n_1 = 100|V|$. The log likelihood can be rewritten as

$$\mathcal{L}(G|F, F') = \sum_{(v_i, v_j) \in E} \log \Pr[(v_i, v_j) \in E] + \sum_{k=1, (v_{ik}, v_{jk}) \notin E}^{n_1} \log \Pr[(v_{ik}, v_{jk}) \notin E]. \quad (11.14)$$

Then the computation of likelihood of network generation only includes $O(E + n_1) = O(V)$ terms.

On the other hand, the computation of Eq. (11.12) takes at least $O(|L|)$ time because the denominator contains $|L|$ terms. Note that the computation of this conditional probability need to be done for every location. Therefore, the computation of trajectory generation needs at least $O(|L|^2)$ which is not efficient. Similarly, we don't compute every term in the denominator. Instead we only compute location $l_i^{(v,j)}$ and other n_2 random locations. In this paper we use $n_2 = 100$. Then we reformulate Eq. (11.12) as

$$\Pr[l_i^{(v,j)} | R_v^{(i,j)}] = \frac{\exp(R_v^{(i,j)} \cdot U'_{l_i^{(v,j)}})}{\exp(R_v^{(i,j)} \cdot U'_{l_i^{(v,j)}}) + \sum_{k=1, l_k \neq l_i^{(v,j)}}^{n_2} \exp(R_v^{(i,j)} \cdot U'_{l_k})}. \quad (11.15)$$

Then the computation of the denominator only includes $O(n_2 + 1) = O(1)$ terms. The parameters are updated with AdaGrad [Duchi et al., 2011], a variant of stochastic gradient descent (SGD), in mini-batches.

Complexity Analysis. We first given the complexity analysis on time cost. The network generation of user v takes $O(d)$ time to compute log likelihood and gradients of F_v and corresponding rows of F'. Thus, the complexity of network generation is $O(d|V|)$. In trajectory generation, we denote the total number of check-in data as $|D|$. Then the forward and backward propagation of GRU take $O(d^2|D|)$ time to compute since the complexity of a single check-in is $O(d^2)$. Each step of RNN takes $O(d^2)$ time to update local dependency representation and compute the gradients of S_0, U, W. The computation of log likelihood and gradients of U', F_v, P_v, S_{i-1}^j, and h^{j-1} takes $O(d^2)$ times. Hence, the overall complexity of our model is $O(d^2|D| + d|V|)$. Note that the representation dimension d and number of negative samples per user/location are much less than the data size $|V|$ and $|D|$. Hence, the time complexity of our algorithm JNTM is linear to the data size and scalable for large datasets. Although the training time complexity of our model is relatively high, the test time complexity is small. When making location recommendations to a user in the test stage, it takes $O(d)$ time to update the hidden states of RNN/LSTM, and $O(d)$ time to evaluate a score for a single location. Usually, the hidden dimensionality d is a small number, which indicates that our algorithm is efficient to make online recommendations.

In terms of space complexity, the network representations F and location representations U take $O((|V| + |L|)d)$ space cost in total. The space cost of other parameters is at most $O(d^2)$, which can be neglected since d is much less than $|V|$ and $|L|$. Thus, the space complexity of our model is similar to that of previous models such as FPMC [Rendle et al., 2010], PRME [Feng et al., 2015], and HRM [Wang et al., 2015].

11.3   EMPIRICAL ANALYSIS

In this section, we evaluate the performance of our proposed model JNTM. We consider two application tasks, namely next-location recommendation and friend recommendation. In what follows, we will discuss the data collection, baselines, parameter setting and evaluation metrics. Then we will present the experimental results together with the related analysis.

11.3.1   DATA COLLECTION

We consider using two publicly available LBSN datasets[3] [Cho et al., 2011], i.e., Gowalla and Brightkite, for our evaluation. Gowalla and Brightkite have released the mobile apps for users. For example, with Brightkite you can track on your friends or any other BrightKite users nearby using a phone's built in GPS; Gowalla has a similar function: use GPS data to show where you are, and what's near you.

These two datasets provide both connection links and users' check-in information. A connection link indicates reciprocal friendship and a check-in record contains the location ID and the corresponding check-in timestamp. We organized the check-in information as trajectory sequences. Following Cheng et al. [2013], we split a trajectory wherever the interval between two successive check-ins is larger than six hours. We preformed some preprocessing steps on both datasets. For *Gowalla*, we removed all users who have less than 10 check-ins and locations which

---

[3]http://snap.stanford.edu/data/

Table 11.2: Statistics of datasets. $|V|$: number of vertices; $|E|$: number of edges; $|D|$: number of check-ins; $|L|$: number of locations.

Dataset	$|V|$	$|E|$	$|D|$	$|L|$
Gowalla	37,800	390,902	2,212,652	58,410
Brightkite	11,498	140,372	1,029,959	51,866

have fewer than 15 check-ins, and finally obtained 837, 352 subtrajectories. For *Brightkite*, since this dataset is smaller, we only remove users who have fewer than 10 check-ins and locations which have fewer than 5 check-ins, and finally obtain 503, 037 subtrajectories after preprocessing. Table 11.2 presents the statistics of the preprocessed datasets. Note that our datasets are larger than those in previous works [Cheng et al., 2013, Feng et al., 2015].

A major assumption we have made is that there exists close association between social links and mobile trajectory behaviors. To verify this assumption, we construct an experiment to reveal basic correlation patterns between these two factors. For each user, we first generate a location set consisting of the locations that have been visited by the user. Then we can measure the similarity degree between the location sets from 2 users using the overlap coefficient.[4] The average overlap coefficients are 11.1% and 15.7% for a random friend pair (i.e., 2 users are social friends) on Brightkite and Gowalla dataset, respectively. As a comparison, the overlap coefficient falls to 0.5% and 0.5% for a random non-friend pair (i.e., 2 users are not social friends) on Brightkite and Gowalla dataset, respectively. This finding indicates that users that are socially connected indeed have more similar visit characteristics. We next examine whether two users with similar trajectory behaviors are more likely to be socially connected. We have found that the probabilities that two random users are social friends are 0.1% and 0.03% on Brightkite and Gowalla dataset, respectively. However, if we select 2 users with more than 3 common locations in their location set, the probabilities that they are social friends increase to 9% and 2%, respectively. The above two findings show social connections are closely correlated with mobile trajectory behaviors in LBSNs.

11.3.2   EVALUATION TASKS AND BASELINES

Next-Location Recommendation: For the task of next-location recommendation, we consider the following baselines.

- **Paragraph Vector (PV)** [Le and Mikolov, 2014] is a representation learning model for both sentence and documents using simple neural network architecture. To model trajectory data, we treat each location as a word and each user as a paragraph of location words.

[4]https://en.wikipedia.org/wiki/Overlap_coefficient

- **Feature-Based Classification (FBC)** solves the next-location recommendation task by casting it as a multi-class classification problem. The user features are learned using DeepWalk algorithm, and the location features are learned using word2vec [Mikolov et al., 2013b] algorithm (similar to the training method of PV above). These features are subsequently incorporated into a a softmax classifier, i.e., a multi-class generalization of logistic regression.

- **FPMC** [Rendle et al., 2010], which is a state-of-the-art recommendation algorithm, factorizes tensor of transition matrices of all users and predicts next location by computing the transition probability based on Markov chain assumption. It was originally proposed for product recommendation, however, it is easy adapt FPMC to deal with next-location recommendation.

- **PRME** [Feng et al., 2015] extends FPMC by modeling user-location and location-location pairs in different vector spaces. PRME achieves state-of-the-art performance on next-location recommendation task.

- **HRM** [Wang et al., 2015] is a latest algorithm for next-basket recommendation. By taking each subtrajectory as a transaction basket, we can easily adapt HRM for next-location recommendation. It is the first study that distributed representation learning has been applied to the recommendation problem.

We select these five baselines, because they represent different recommendation algorithms. PV is based on simple neural networks, FBC is a traditional classification model using embedding features, FPMC is mainly developed in the matrix factorization framework, PRME makes specific extensions based on FPMC to adapt to the task of next-location recommendation, and HRM adopts the distributed representation learning method for next-basket modeling.

Next, we split the data collection into the training set and test set. The first 90% of check-in subtrajectories of each user are used as the training data and the remaining 10% as test data. To tune the parameters, we use the last 10% of check-ins of training data as the validation set.

Given a user, we predict the locations in the test set in a sequential way: for each location slot, we recommend five or ten locations to the user. For JNTM, we naturally rank the locations by the log likelihood as shown in Eq. (11.12). Note that negative sampling is not used in evaluation. For the baselines, we rank the locations by the transition probability for FPMC and HRM and transition distance for PRME. The predictions of PV and FBC can be obtained from the output of softmax layer of their algorithms. Then we report Recall@5 and Recall@10 as the evaluation metrics.

Friend Recommendation: For the task of friend recommendation, we consider three kinds of baselines based on the used data resources, including the method with only the networking data (i.e., DeepWalk), the method with only the trajectory data (i.e., PMF), and the methods with both networking and trajectory data (i.e., PTE and TADW).

- **DeepWalk** [Perozzi et al., 2014] is a state-of-the-art NRL method which learns vertex embeddings from random walk sequences. It first employs the random walk algorithm to generate length-truncated random paths, and apply the word embedding technique to learn the representations for network vertices.

- **PMF** [Mnih and Salakhutdinov, 2007] is a general collaborative filtering method based on user-item matrix factorization. In our experiments, we build the user-location matrix using the trajectory data, and then we utilize the user latent representations for friend recommendation.

- **PTE** [Tang et al., 2015a] develops a semi-supervised text embedding algorithm for unsupervised embedding learning by removing the supervised part and optimizing over adjacency matrix and user-location co-occurrence matrix. PTE models a conditional probability $p(v_j|v_i)$ which indicates the probability that a given neighbor of v_i is v_j. We compute the conditional probabilities for friend recommendation.

- **TADW** [Yang et al., 2015] further extends DeepWalk to take advantage of text information of a network. We can replace text feature matrix in TADW with user-location co-occurrence matrix by disregarding the sequential information of locations. TADW defines an affinity matrix where each entry of the matrix characterizes the strength of the relationship between corresponding users. We use the corresponding entries of affinity matrix to rank candidate users for recommendation.

To construct the evaluation collection, we randomly select 20–50 of the existing connection links as training set and leave the rest for test. We recommend 5 or 10 friends for each user and report Recall@5 and Recall@10. The final results are compared by varying the training ratio from 20–50%. Specifically, for each user v, we take all the other users who are not her friends in the training set as the candidate users. Then, we rank the candidate users, and recommend top 5 or 10 users with highest ranking scores. To obtain the ranking score of user v_j when we recommend friends for user v_i, DeepWalk and PMF adopt the cosine similarity between their user representations. For PTE, we use the conditional probability $p(v_j|v_i)$ which indicates the probability that a given neighbor of v_i is v_j as ranking scores. For TADW, we compute the affinity matrix A and use the corresponding entry A_{ij} as ranking scores. For our model, we rank users with highest log likelihood according to Eq. (11.1).

The baselines methods and our model involves an important parameter, i.e., the number of latent (or embedding) dimensions. We use a grid search from 25–100 and set the optimal value using the validation set. Other parameters in baselines or our model can be tuned in a similar way. For our model, the learning rate and number of negative samples are empirically set to

Table 11.3: Results of different methods on next location recommendation

Dataset	Brightkite			Gowalla		
Metric (%)	R@1	R@5	R@10	R@1	R@5	R@10
PV	18.5	44.3	53.2	9.9	27.8	36.3
FBC	16.7	44.1	54.2	13.3	34.4	42.3
FPMC	20.6	45.6	53.8	10.1	24.9	31.6
PRME	15.4	44.6	53.0	12.2	31.9	38.2
HRM	17.4	46.2	56.4	7.4	26.2	37.0
JNTM	**22.1**	**51.1**	**60.3**	**15.4**	**38.8**	**48.1**

0.1 and 100, respectively. We randomly initialize parameters according to uniform distribution $U(-0.02, 0.02)$.

11.3.3   EXPERIMENTAL RESULTS ON NEXT-LOCATION RECOMMENDATION

Table 11.3 shows the results of different methods on next-location recommendation. Compared with FPMC and PRME, HRM models the sequential relatedness between consecutive subtrajectories while the sequential relatedness in a subtrajectory is ignored. In the Brightkite dataset, the average number of locations in a subtrajectory is much less than that in the Gowalla dataset. Therefore, short-term sequential contexts are more important in the Gowalla dataset and less useful in the Brightkite dataset. Experimental results in Table 11.3 demonstrate this intuition: HRM outperforms FPMC and PRME on Brightkite while PRME works best on Gowalla.

As shown in Table 11.3, our model JNTM consistently outperforms the other baseline methods. JNTM yields 4.9% and 4.4% improvement on Recall@5 as compared to the best baseline HRM on the Brightkite dataset and FBC on the Gowalla dataset. Recall that our model JNTM has considered four factors, including user preference, influence of friends, and short-term and long-term sequential contexts. All the baseline methods only characterize user preference (or friend influence for FBC) and a single kind of sequential contexts. Thus, JNTM achieves the best performance on both datasets.

The above results are reported by averaging over all the users. In recommender systems, an important issue is how a method performs in the cold-start setting, i.e., new users or new items. To examine the effectiveness on new users generating very few check-ins, we present the results of Recall@5 for users with no more five subtrajectories in Table 11.4. In a cold-start scenario, a commonly used way to leverage the side information (e.g., user links [Cheng et al., 2012] and text information [Gao et al., 2015a, Li et al., 2010, Zhao et al., 2015a]) to alleviate the data sparsity. For our model, we characterize two kinds of user representations, either using network data or trajectory data. The user representations learned using network data can be exploited

Table 11.4: Results of next location recommendation results for users with no more than five subtrajectories

Dataset	Brightkite			Gowalla		
Metric (%)	R@1	R@5	R@10	R@1	R@5	R@10
PV	13.2	22.0	26.1	4.6	7.8	9.2
FBC	9.0	29.6	39.5	4.9	12.0	16.3
FPMC	17.1	30.0	33.9	5.5	13.5	18.5
PRME	22.4	36.3	40.0	7.2	12.2	15.1
HRM	12.9	31.2	39.7	5.2	15.2	21.5
JNTM	**28.4**	**53.7**	**59.3**	**10.2**	**24.8**	**32.0**

Table 11.5: Results of different methods on next new location recommendation

Dataset	Brightkite			Gowalla		
Metric (%)	R@1	R@5	R@10	R@1	R@5	R@10
PV	0.5	1.5	2.3	1.0	3.3	5.3
FBC	0.5	1.9	3.0	1.0	3.1	5.1
FPMC	0.8	2.7	4.3	2.0	6.2	9.9
PRME	0.3	1.1	1.9	0.6	2.0	3.3
HRM	1.2	3.5	5.2	1.7	5.3	8.2
JNTM	**1.3**	**3.7**	**5.5**	**2.7**	**8.1**	**12.1**

to improve the recommendation performance for new users to some extent. By utilizing the networking representations, the results indicate that our model JNTM is very promising to deal with next-location recommendation in a cold-start setting.

Note that the above experiments are based on general next-location recommendation, where we do not examine whether a recommended location has been previously visited or not by a user. To further test the effectiveness of our algorithm, we conduct experiments on next new location recommendation task proposed by previous studies [Feng et al., 2015]. In this setting, we only recommend new locations when the user decide to visit a place. Specifically, we rank all the locations that a user has never visited before for recommendation [Feng et al., 2015]. We present the experimental results in Table 11.5. Our method consistently outperforms all the baselines on next new location recommendation in both datasets. By combining results in Tables 11.3 and 11.4, we can see that our model JNTM is more effective in next-location recommendation task compared to these baselines.

Table 11.6: Performance comparison for three variants of JNTM on next-location recommendation

Dataset	Brightkite			Gowalla		
Metric (%)	R@1	R@5	R@10	R@1	R@5	R@10
JNTM$_{base}$	20.2	49.3	59.2	12.6	36.6	45.5
JNTM$_{base+long}$	20.4 (+2%)	50.2 (+2%)	59.8 (+1%)	13.9 (+10%)	36.7 (+0%)	45.6 (+0%)
JNTM$_{base+long+short}$	**22.1** (+9%)	**51.1** (+4%)	**60.3** (+2%)	**15.4** (+18%)	**38.8** (+6%)	**48.1** (+6%)

Table 11.7: Performance comparison for three variants of JNTM on next new location recommendation

Dataset	Brightkite			Gowalla		
Metric (%)	R@1	R@5	R@10	R@1	R@5	R@10
JNTM$_{base}$	0.8	2.5	3.9	0.9	3.3	5.5
JNTM$_{base+long}$	1.0 (+20%)	3.3 (+32%)	4.8 (+23%)	1.0 (+11%)	3.5 (+6%)	5.8 (+5%)
JNTM$_{base+long+short}$	**1.3** (+63%)	**3.7** (+48%)	**5.5** (+41%)	**2.7** (+200%)	**8.1** (+145%)	**12.1** (+120%)

In the above, we have shown the effectiveness of the proposed model JNTM on the task of next-location recommendation. Since trajectory data itself is sequential data, our model has leveraged the flexibility of recurrent neural networks for modeling sequential data, including both short-term and long-term sequential contexts. Now we study the effect of sequential modeling on the current task.

We prepare three variants for our model JNTM.

- JNTM$_{base}$: it removes both short-term and long-term contexts. It only employs the user interest representation and network representation to generate the trajectory data.

- JNTM$_{base+long}$: it incorporates the modeling for long-term contexts to JNTM$_{base}$.

- JNTM$_{base+long+short}$: it incorporates the modeling for both short-term and long-term contexts to JNTM$_{base}$.

Tables 11.6 and 11.7 show the experimental results of three JNTM variants on the Brightkite and Gowalla dataset. The numbers in the brackets indicate the relative improvement against JNTM$_{base}$. We can observe a performance ranking: JNTM$_{base}$ < JNTM$_{base+long}$ < JNTM$_{base+long+short}$. The observations indicate that both kinds of sequential contexts are useful to improve the performance for next-location recommendation. In general, next location recommendation (i.e., both old and new locations are considered for recommendation), we can see

Table 11.8: Friend recommendation results on Brightkite dataset

Training ratio	20%		30%		40%		50%	
Metric (%)	R@5	R@10	R@5	R@10	R@5	R@10	R@5	R@10
DeepWalk	2.3	3.8	3.9	6.7	5.5	9.2	7.4	12.3
PMF	2.1	3.6	2.1	3.7	2.3	3.4	2.3	3.8
PTE	1.5	2.5	3.8	4.7	4.0	6.6	5.1	8.3
TADW	2.2	3.4	3.6	3.9	2.9	4.3	3.2	4.5
JNTM	3.7	6.0	5.4	8.7	6.7	11.1	8.4	13.9

Table 11.9: Friend recommendation results on Gowalla dataset

Training ratio	20%		30%		40%		50%	
Metric (%)	R@5	R@10	R@5	R@10	R@5	R@10	R@5	R@10
DeepWalk	2.6	3.9	5.1	8.1	7.9	12.1	10.5	15.8
PMF	1.7	2.4	1.8	2.5	1.9	2.7	1.9	3.1
PTE	1.1	1.8	2.3	3.6	3.6	5.6	4.9	7.6
TADW	2.1	3.1	2.6	3.9	3.2	4.7	3.6	5.4
JNTM	3.8	5.5	5.9	8.9	7.9	11.9	10.0	15.1

that the improvement from short- and long-term context is not significant. The explanation is that a user is likely to show repeated visit behaviors (e.g., visiting the locations that have been visited before), and thus user preference is more important than sequential context to improve the recommendation performance. While for next new location recommendation, the sequential context especially short-term context yields a large improvement margin over the baseline. These results indicate that the sequential influence is more important than user preference for new location recommendation. Our finding is also consistent with previous work [Feng et al., 2015], i.e., sequential context is important to consider for next new location recommendation.

11.3.4 EXPERIMENTAL RESULTS ON FRIEND RECOMMENDATION

We continue to present and analyze the experimental results on the task of friend recommendation. Tables 11.9 and 11.8 show the results when the training ratio varies from 20–50%.

Among the baselines, DeepWalk performs best and even better than the baselines using both networking data and trajectory data (i.e., PTE and TADW). A major reason is that DeepWalk is tailored to the reconstruction of network connections and adopts a distributed representation method to capture the topology structure. Although PTE and TADW utilize both network and trajectory data, their performance is still low. These two methods cannot capture

Table 11.10: Friend recommendation results for users with fewer than five friends when training ratio is 50%

Dataset	Brightkite		Gowalla	
Metric (%)	R@5	R@10	R@5	R@10
DeepWalk	14.0	18.6	19.8	23.5
JNTM	**16.1**	**20.4**	**21.3**	**25.5**

the sequential relatedness in trajectory sequences. Another observation is that PMF (i.e., factorizing the user-location matrix) is better than PTE at the ratio of 20% but becomes the worst baseline. It is because that PMF learns user representations using the trajectory data, and the labeled data (i.e., links) is mainly used for training a classifier.

Our algorithm is competitive with state-of-the-art network embedding method Deep-Walk and outperforms DeepWalk when network structure is sparse. The explanation is that trajectory information is more useful when network information is insufficient. As network becomes dense, the trajectory information is not as useful as the connection links. To demonstrate this explanation, we further report the results for users with fewer than five friends when the training ratio of 50%. As shown in Table 11.10, our methods have yielded 2.1% and 1.5% improvement than DeepWalk for these inactive users on the Brightkite and Gowalla datasets, respectively. The results indicate that trajectory information is useful to improve the performance of friend recommendation for users with very few friends.

In summary, our methods significantly outperforms previous methods on both next-location prediction and friend recommendation. Experimental results on both tasks demonstrate the effectiveness of our proposed model.

11.4 FURTHER READING

In this chapter, we focus on recommendation, systems on a specific type of networks, i.e., LB-SNs [Bao et al., 2015]. A typical application task in LBSN is the location recommendation, which aims to infer users' visit preference and make meaningful recommendations for users to visit. It can be divided into three different settings: general location recommendation, time-aware location recommendation, and next-location recommendation. General location recommendation will generate an overall recommendation list of locations for a users to visit, while time-aware or next location recommendation further imposes the temporal constraint on the recommendation task by either specifying the time period or producing sequential predictions.

For general location recommendation, several kinds of side information are considered, such as geographical [Cheng et al., 2012, Ye et al., 2011], temporal [Zhao et al., 2016], and social network information [Levandoski et al., 2012]. To address the data sparsity issue, content information including location category labels is also concerned [Yin et al., 2013, Zhou et al.,

2016]. The location labels and tags can also be used in probabilistic model such as aggregate LDA [Gao et al., 2015a]. Textual information which includes text descriptions [Gao et al., 2015a, Li et al., 2010, Zhao et al., 2015a] are applied for location recommendation as well. W^4 employs tensor factorization on multi-dimensional collaborative recommendation for Who (user), What (location category), When (time), and Where (location) [Bhargava et al., 2015, Zheng et al., 2010].

For time-aware location recommendation task which recommends locations at a specific time, it is also worth modeling the temporal effect. Collaborate filtering-based method [Yuan et al., 2013] unifies temporal and geographical information with linear combination. Geographical-temporal graph was proposed for time-aware location recommendation by doing preference propagation on the graph [Yuan et al., 2014b]. In addition, temporal effect is also studied via nonnegative matrix factorization [Gao et al., 2013], embedding [Xie et al., 2016], and RNN [Liu et al., 2016].

Different from general location recommendation, next-location recommendation also need to take current state into account. Therefore, the sequential information is more important to consider in next location recommendation. Most previous works model sequential behaviors, i.e., trajectories of check-in locations, based on Markov chain assumption which assumes the next location is determined only by current location and independent of previous ones [Cheng et al., 2013, Rendle et al., 2010, Ye et al., 2013, Zhang et al., 2014]. For example, Factorized Personalized Markov Chain (FPMC) algorithm [Rendle et al., 2010] factorizes the tensor of transition cube which includes transition probability matrices of all users. Personalized Ranking Metric Embedding (PRME) [Feng et al., 2015] further extends FPMC by modeling user-location distance and location-location distance in two different vector spaces. These methods are applied for next-location recommendation which aims at predicting the next location that a user will visit, given check-in history and current location of the user. Note that Markov chain property is a strong assumption that assumes next location is determined only by current location. In practice, next location may also be influenced by the entire check-in history. Recently, RNN-based methods [Gao et al., 2017a, Li et al., 2020] have demonstrated their effectiveness in recommendation on LBSNs.

Next, we would like to introduce some recent advances based on graph embedding techniques in more general recommendation scenarios.

GNNs have been recognized as a powerful technique to encode graphs into low-dimensional vector representations. sRMGCNN [Monti et al., 2017] and GCMC [van den Berg et al., 2017] first leverage GNNs in recommendation tasks and achieve state-of-the-art results.

sRMGCNN constructs k-nearest neighbor graphs for users and items to extract their feature vectors. The features are input into a recurrent neural network to recover the rating matrix. GCMC is based on the GCN [Kipf and Welling, 2017], which aggregates and updates directly in the spatial domain of the graph. GCMC can be interpreted as an encoder-decoder structure. The model first obtains the representation vectors of the user and item nodes through the graph encoder, and then predicts the ratings through a bilinear decoder. Due to the representation power of GCN, GCMC has achieved state-of-the-art results on multiple real-world benchmarks.

PinSage [Ying et al., 2018] is a web-scale GNN-based recommendation algorithm which designs an efficient computational pipeline for Pinterest item recommendation. Although being effective, PinSage depends on node textual or visual features heavily, while they are not available in many benchmarks. STAR-GCN [Zhang et al., 2019b] employs a multi-block graph auto-encoder architecture. The intermediate reconstruction supervision boosts the model performance consistently.

For other recommendation scenarios, NGCF [Wang et al., 2019c] applies GNNs on sequential recommendation, GraphRec [Fan et al., 2019] focuses on the problem of social recommendation and adopts GAT [Veličković et al., 2018] to capture the importance of neighbors, where users are associated with social relationships. KGAT [Wang et al., 2019b] focuses on KG-enhanced personalized recommendation by constructing the user-item-entity graph.

Part of this chapter was published in our journal paper by Yang et al. [2017b] on ACM TOIS.

CHAPTER 12

Network Embedding for Information Diffusion Prediction

The study of information diffusion or cascade has attracted much attention over the last decade. Most related works target on studying cascade-level *macroscopic* properties such as the final size of a cascade. Existing *microscopic* cascade models which focus on user-level modeling either make strong assumptions on how a user gets infected by a cascade or limit themselves to a specific scenario where "who infected whom" information is explicitly labeled. The strong assumptions oversimplify the complex diffusion mechanism and prevent these models from better fitting real-world cascade data. Also, the methods which focus on specific scenarios cannot be generalized to a general setting where the diffusion graph is unobserved. To overcome the drawbacks of previous works, we propose a Neural Diffusion Model (NDM) for general microscopic cascade study. NDM makes relaxed assumptions and employs deep learning techniques to characterize user embeddings in a cascade. NE methods are also employed to utilize the underlying social network information and can provide additional auxiliary inputs for the model. These advantages enable our model to go beyond the limitations of previous methods, better fit the diffusion data and generalize to unseen cascades. Experimental results on diffusion prediction task over four realistic cascade datasets show that our model can achieve a relative improvement up to 26% against the best performing baseline in terms of F1 score.

12.1 OVERVIEW

Information diffusion is a ubiquitous and fundamental event in our daily lives, such as the spread of rumors, contagion of viruses and propagation of new ideas and technologies. The diffusion process, also called a *cascade*, has been studied over a broad range of domains. Though some works believe that even the eventual size of a cascade cannot be predicted [Salganik et al., 2006], recent works [Cheng et al., 2014, Yu et al., 2015, Zhao et al., 2015b] have shown the ability to estimate the size, growth, and many other key properties of a cascade. Nowadays, the modeling of cascades play an important role in many real-world applications, e.g., production recommendation [Aral and Walker, 2012, Domingos and Richardson, 2001, Leskovec et al., 2006, 2007, Watts and Dodds, 2007], epidemiology [Hethcote, 2000, Wallinga and Teunis, 2004], social

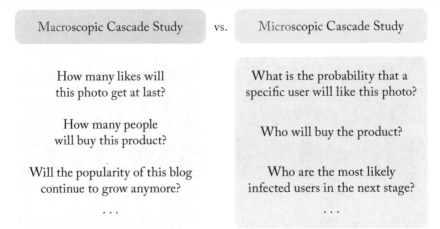

Figure 12.1: Macroscopic cascade study vs. microscopic cascade study.

networks [Dow et al., 2013, Kempe et al., 2003, Lappas et al., 2010], and the spread of news and opinions [Gruhl et al., 2004, Leskovec et al., 2009, Liben-Nowell and Kleinberg, 2008]. Most previous works focus on the study of *macroscopic* properties such as the total number of users who share a specific photo [Cheng et al., 2014] and the growth curve of the popularity of a blog [Yu et al., 2015]. However, macroscopic cascade study is a rough estimate of cascades and cannot be adapted for *microscopic* questions, as shown in Fig. 12.1. Microscopic cascade study, which pays more attention to user-level modeling instead of cascade-level, is much more powerful than macroscopic estimation and allows us to apply user-specific strategies for real-world applications. For example, during the adoption of a new product, microscopic cascade study can help us deliver advertisements to those users that are most likely to buy the product at each stage. In this paper, we focus on the study of microscopic level.

Though useful and powerful, the microscopic study of cascades faces great challenges because the real-world diffusion process could be rather complex [Romero et al., 2011] and usually partially observed [Myers and Leskovec, 2010, Wallinga and Teunis, 2004]:

Complex mechanism. Since the mechanism of how a specific user gets infected[1] is sophisticated, traditional cascade models based on strong assumptions and simple formulas may not be the best choice for microscopic cascade modeling. Existing cascade models [Bourigault et al., 2016, Gomez-Rodriguez et al., 2012, Gomez Rodriguez et al., 2013, Rodriguez et al., 2014] which could be adopted for microscopic analysis mostly ground in Independent Cascade (IC) model [Kempe et al., 2003]. IC model assigns a static probability $p_{u,v}$ to user pairs (u, v) with pairwise independent assumptions, where the probability $p_{u,v}$ indicates how likely user v will get infected by user u when u is infected. Other diffusion models [Bourigault et al., 2014,

---

[1]We use "infected" and "activated" alternatively to indicate that a user is influenced by a cascade.

Gao et al., 2017b] make even stronger assumptions that the infected users are only determined by the source user. Though intuitive and easy to understand, these cascade models are based on strong assumptions and oversimplified probability estimation formulas, both of which limit the expressivity and ability to fit complex real-world cascade data [Li et al., 2017a]. The complex mechanism of real-world diffusions encourages us to explore more sophisticated models, e.g., deep learning techniques, for cascade modeling.

Incomplete observation. On the other hand, the cascade data is usually partially observed indicates that we can only observe those users getting infected without knowing who infected them. However, to the best of our knowledge, existing deep-learning engined microscopic cascade models [Hu et al., 2018, Wang et al., 2017c] are based on the assumption that the diffusion graph where a user can only infect and get infected by its neighbors is already known. For example, when we study the retweeting behavior on the Twitter network, "who infected whom" information is explicitly labeled in retweet chain and the next infected user candidates are restricted to the neighboring users rather than the whole user set. While in most diffusion processes such as the adoption of a product or the contamination of a virus, the diffusion graph is unobserved [Myers and Leskovec, 2010, Wallinga and Teunis, 2004, Zekarias Kefato and Montresor, 2017]. Therefore, these methods consider a much simpler problem and cannot be generalized to a general setting where the diffusion graph is unknown.

To fill in the blank of general microscopic cascade study and address the limitations of traditional cascade models, we propose a neural diffusion model based on relaxed assumptions and employ up-to-date deep learning techniques, i.e., attention mechanism and convolutional neural network, for cascade modeling. Network embedding methods are also employed to utilize the underlying social network information. The relaxed assumptions enable our model to be more flexible and less constrained, and deep learning tools are good at capturing the complex and intrinsic relationships that are hard to be characterized by hand-crafted features. Both advantages allow our model to go beyond the limitations of traditional methods based on strong assumptions and oversimplified formulas and better fit the complicated cascade data. Following the experimental settings in Bourigault et al. [2016], we conduct experiments on diffusion prediction task over four realistic cascade datasets to evaluate the performances of our proposed model and other state-of-the-art baseline methods. Experimental results show that our model can achieve a relative improvement up to 26% against the best performing baseline in terms of F1 score.

12.2   METHOD: NEURAL DIFFUSION MODEL (NDM)

In this section, we will start by formalizing the problem and introducing the notations. Then we propose two heuristic assumptions according to the data observations as our basis and design an NDM using deep learning techniques. Finally, we will introduce the overall optimization function and other details of our model.

12.2.1 PROBLEM FORMALIZATION

A cascade dataset records the information that an item spreads to whom and when during its diffusion. For example, the item could be a product and the cascade records who bought the product at what moment. However, in most cases, there exists no explicit interaction graph between the users [Bourigault et al., 2016, Saito et al., 2008]. Therefore, we have no explicit information about how a user was infected by other users.

Formally, given user set $\mathcal{U}$ and observed cascade sequence set $\mathcal{C}$, each cascade $c_i \in \mathcal{C}$ consists a list of users $\{u_0^i, u_1^i \ldots u_{|c_i|-1}^i\}$ ranked by their infection time, where $|c_i|$ is the length of sequence c_i and $u_j^i \in \mathcal{U}$ is the j-th user in the sequence c_i. Note that we only consider the order of users getting infected and ignore the exact timestamps of infections as previous works did [Bourigault et al., 2016, Wang et al., 2017c, Zekarias Kefato and Montresor, 2017].

In this paper, our goal is to learn a cascade model which can predict the next infected user u_{j+1} given a partially observed cascade sequence $\{u_0, u_1 \ldots u_j\}$. The learned model is able to predict the entire infected user sequence based on the first few observed infected users and thus be used for microscopic evaluation tasks illustrated in Fig. 12.1. In our model, we add a virtual user called "Terminate" to the user set $\mathcal{U}$. At training phase, we append "Terminate" to the end of each cascade sequence and allow the model to predict next infected user as "Terminate" to indicate that no more users will be infected in this cascade.

Further, we represent each user by a parameterized real-valued vector to project users into vector space. The real-valued vectors are also called embeddings. We denote the embedding of user u as $emb(u) \in \mathbb{R}^d$ where d is the dimension of embeddings. In our model, a larger inner product between the embeddings of two users indicates a stronger correlation between the users. The embedding layer is used as the bottom layer of our model by projecting a user into corresponding vector as shown in Fig. 12.2.

12.2.2 MODEL ASSUMPTIONS

In the traditional IC model [Kempe et al., 2003] settings, all previously infected users can activate a new user independently and equally regardless of their orders of getting infected. Many extensions of IC model further considered time delay information such as continuous time IC (CTIC) [Saito et al., 2009] and Netrate [Rodriguez et al., 2014]. However, none of these models tried to find out which users are actually active and more likely to activate other users at the moment. To address this issue, we propose the following assumption.

Assumption 1. Given a recently infected user u, users that are strongly correlated to user u including user u itself are more likely to be active.

This assumption is intuitive and straightforward. As a newly activated user, u should be active and may infect other users. The users strongly correlated to user u are probably the reason why user u gets activated recently and thus more likely to be active than other users at the

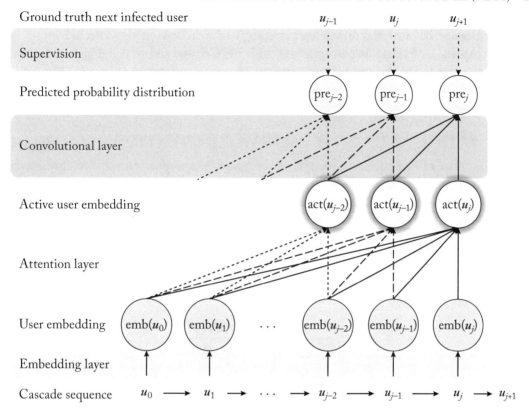

Figure 12.2: An overview of our Neural Diffusion Model (NDM). NDM sequentially predicts the next infected user based on the active embeddings (the blue nodes) of recent activated users and the active embeddings is computed by an attention layer over the user embeddings (the green nodes) of all previous infected users.

moment. We further propose the concept of "active user embedding" to characterize all such active users.

Definition 1. For each recently infected user u, we aim to learn an active user embedding $act(u) \in \mathbb{R}^d$ which represents the embedding of all active users related to user u, and can be used for predicting the next infected user in next step.

The active user embedding $act(u_j)$ characterizes the potential active users related to the fact that user u_j gets infected. From the data observations, we can see that all recently infected users could be relevant to the next infected one. Therefore, the active user embeddings of all recently infected users should contribute to the prediction of next infected user, which leads to the following assumption.

Assumption 2. All recently infected users should contribute to the prediction of next infected user and be processed differently according to the order of getting infected.

Compared with the strong assumptions made by IC-based and embedding-based method introduced in related works, our heuristic assumptions allow our model to be more flexible and better fit cascade data. Now we will introduce how to build our model based on these two assumptions, i.e., extracting active users and unifying these embeddings for prediction.

12.2.3   EXTRACTING ACTIVE USERS WITH ATTENTION MECHANISM

For the purpose of computing active user embeddings, we propose to use attention mechanism [Bahdanau et al., 2015, Vaswani et al., 2017] to extract the most likely active users by giving them more weights than other users. As shown in Fig. 12.2, the active embedding of user u_j is computed as a weighted sum of previously infected users:

$$act(u_j) = \sum_{k=0}^{j} w_{jk} emb(u_k), \tag{12.1}$$

where the weight of u_k is

$$w_{jk} = \frac{\exp(emb(u_j)emb(u_k)^T)}{\sum_{m=0}^{j} \exp(emb(u_j)emb(u_m)^T)}. \tag{12.2}$$

Note that $w_{jk} \in (0, 1)$ for every k and $\sum_{m=0}^{j} w_{jm} = 1$. w_{jk} is the normalized inner product between the embeddings of u_j and u_k which indicates the strength of correlation between them.

From the definition of active user embedding $act(u_j)$ in Eq. (12.1), we can see that the user embeddings $emb(u_k)$ which have a larger inner product with $emb(u_j)$ will be allocated a larger weight w_{jk}. This formula naturally follows our assumption that users strongly correlated to user u including user u itself should be paid more attention.

To fully utilize the advantages of a neural model, we further employ the multi-head attention [Vaswani et al., 2017] to improve the expressibility. Multi-head attention projects the user embeddings into multiple subspaces with different linear projections. Then multi-head attention performs attention mechanism on each subspace independently. Finally, multi-head attention concatenates the attention embeddings in all subspaces and feeds the result into a linear projection again.

Formally, in a multi-head attention with h heads, the embedding of i-th head is computed as

$$head_i = \sum_{k=0}^{j} w_{jk}^i emb(u_k) W_i^V, \tag{12.3}$$

where

$$w_{jk}^i = \frac{\exp(emb(u_j)W_i^Q(emb(u_k)W_i^K)^T)}{\sum_{m=0}^{j} \exp(emb(u_j)W_i^Q(emb(u_m)W_i^K)^T)}, \tag{12.4}$$

$W_i^V, W_i^Q, W_i^K \in \mathbb{R}^{d \times d}$ are head-specific linear projection matrices. In particular, W_i^Q and W_i^K can be seen to project user embeddings into *receiver* space and *sender* space, respectively, for asymmetric modeling.

Then we have the active user embedding $act(u_j)$

$$act(u_j) = [head_1, head_2 \dots head_h]W^O, \tag{12.5}$$

where [] indicates concatenation operation and $W^O \in \mathbb{R}^{hd \times d}$ projects the concatenated results into d-dimensional vector space.

Multi-head attention allows the model to "divide and conquer" information from different perspectives (i.e., subspaces) independently and thus is more powerful than the traditional attention mechanism.

12.2.4   UNIFYING ACTIVE USER EMBEDDINGS FOR PREDICTION WITH CONVOLUTIONAL NETWORK

Different from previous works [Rodriguez et al., 2011, 2014] which directly give a time-decay weight that assumes larger weights for the most recently infected users, we propose to use a parameterized neural network to handle the active user embeddings at different positions. Compared with a predefined exponential-decay weighting function [Rodriguez et al., 2014], a parameterized neural network can be learned automatically to fit the real-world dataset and capture the intrinsic relationship between active user embedding at each position and next infected user prediction. In this paper, we consider CNN to meet this purpose.

CNN has been widely used in image recognition [LeCun et al., 2015], recommender systems [Van den Oord et al., 2013], and natural language processing [Collobert and Weston, 2008]. CNN is a *shift-invariant* neural network and allows us to assign position-specific linear projections to the embeddings.

Figure 12.2 illustrates an example where the window size of our convolutional layer $win = 3$. The convolutional layer first converts each active user embedding $act(u_{j-n})$ into a $|\mathcal{U}|$-dimensional vector by a position-specific linear projection matrix $W_n^C \in \mathbb{R}^{d \times |\mathcal{U}|}$ for $n = 0, 1 \dots win - 1$. Then the convolutional layer sums up the projected vectors and normalizes the summation by *softmax* function.

Formally, given partially observed cascade sequence $(u_0, u_1 \dots u_j)$, the predicted probability distribution $pre_j \in \mathbb{R}^{|\mathcal{U}|}$ is

$$pre_j = \text{softmax}\left(\sum_{n=0}^{win-1} act(u_{j-n})W_n^C\right), \tag{12.6}$$

where $\text{softmax}(x)[i] = \frac{\exp(x[i])}{\sum_p \exp(x[p])}$ and $x[i]$ denotes the i-th entry of a vector x. Each entry of pre_j represents the probability that the corresponding user gets infected at next step.

Since the initial user u_0 plays an important role in the whole diffusion process, we further take u_0 into consideration:

$$pre_j = \text{softmax}(\sum_{n=0}^{win-1} act(u_{j-n})W_n^C + act(u_0)W_{init}^C \cdot F_{init}),\qquad(12.7)$$

where $W_{init}^C \in \mathbb{R}^{d \times |\mathcal{U}|}$ is the projection matrix for initial user u_0 and $F_{init} \in \{0, 1\}$ is a hyperparameter which controls whether incorporate initial user for prediction or not.

12.2.5   OVERALL ARCHITECTURE, MODEL DETAILS, AND LEARNING ALGORITHMS

We naturally maximize the log-likelihood of all observed cascade sequences to build the overall optimization function:

$$\mathcal{L}(\Theta) = \sum_{c_i \in \mathcal{C}} \sum_{j=0}^{|c_i|-2} \log pre_j^i[u_{j+1}^i],\qquad(12.8)$$

where $pre_j^i[u_{j+1}^i]$ is the predicted probability of ground truth next infected user u_{j+1}^i at position j in cascade c_i, and Θ is the set of all parameters need to be learned, including user embeddings $emb(u) \in \mathbb{R}^d$ for each $u \in \mathcal{U}$, projection matrices in multi-head attention $W_n^V, W_n^Q, W_n^K \in \mathbb{R}^{d \times d}$ for $n = 1, 2 \ldots h$, $W^O \in \mathbb{R}^{hd \times d}$ and projection matrices in convolutional layer $W_{init}^C, W_n^C \in \mathbb{R}^{d \times |\mathcal{U}|}$ for $n = 0, 1 \ldots win - 1$. Note that our model is general and can also be adapted for the case where "who infected whom" info is provided by converting a training cascade into a set of "true" infection sequences where each user is infected by its precedent user.

Complexity. The space complexity of our model is $O(d|\mathcal{U}|)$ where d is the embedding dimension which is much less than the size of user set. Note that the space complexity of training traditional IC model will go up to $O(|\mathcal{U}|^2)$ because we need to assign an infection probability between each pair of potential linked users. Therefore, the space complexity of our neural model is less than that of traditional IC methods.

The computation of a single active embedding takes $O(|c_i|d^2)$ time where c_i is the length of corresponding cascade and the next infected user prediction in Eq. (12.7) step takes $O(d|\mathcal{U}|)$ time. Hence, the time complexity of training a single cascade is $O(\sum_{c_i \in \mathcal{C}}(|c_i|^2 d^2 + |c_i|d|\mathcal{U}|))$ which is competitive with previous neural-based models such as embedded IC model [Bourigault et al., 2016]. But our model converges much faster than embedded IC model and is capable of handling large-scale dataset.

12.3   EMPIRICAL ANALYSIS

We conduct experiments on microscopic diffusion prediction task as previous works did [Bourigault et al., 2016] to evaluate the performance of our model and various baseline methods. We will first introduce the datasets, baseline methods, evaluation metrics, and hyperparameter settings. Then we will present the experimental results and give further analysis about the evaluation.

12.3.1   DATASETS

We collect four real-world cascade datasets that cover a variety of applications for evaluation. A cascade is an item or some kind of information that spreads through a set of users. Each cascade consists of a list of *(user,timestamp)* pairs where each pair indicates the fact that the user gets infected at the timestamp.

Lastfm is a music streaming website. We collect the dataset from Celma [2010]. The dataset contains the full history of nearly 1,000 users and the songs they listened to over one year. We treat each song as an item spreading through users and remove the users who listen to no more than five songs.

Irvine is an online community for students at University of California, Irvine collected from Opsahl and Panzarasa [2009]. Students can participate in and write posts on different forums. We regard each forum as an information item and remove the users who participate in no more than five forums.

Memetracker[2] collects a million of news stories and blog posts and track the most frequent quotes and phrases, i.e., memes, for studying the migration of memes across a group of people. Each meme is considered to be an information item and each URL of websites or blogs is regarded as a user. Following the settings of previous works [Bourigault et al., 2016], we filter the URLs to only keep the most active ones to alleviate the effect of noise.

Twitter dataset [Hodas and Lerman, 2014] concerns tweets containing URLs posted on Twitter during October 2010. The complete tweeting history of each URL is collected. We consider each distinct URL as a spreading item over Twitter users. We filter out the users with no more than five tweets. Note that the scale of Twitter dataset is competitive and even larger than the datasets used in previous neural-based cascade modeling algorithms [Bourigault et al., 2016, Wang et al., 2017c].

Note that all the above datasets have no explicit evidence about by whom a user gets infected. Though we have the following relationship in Twitter dataset, we still cannot trace the source of by whom a user is encouraged to tweet a specific URL unless the user directly retweets.

We list the statistics of datasets in Table 12.1. Since we have no interaction graph information between users, we assume that there exists a link between two users if they appear in the same cascade sequence. Each virtual "link" will be assigned a parameterized probability in traditional IC model and thus the space complexity of traditional methods is relatively high

---

[2]http://www.memetracker.org

Table 12.1: Statistics of datasets

Dataset	# Users	# Links	# Cascades	Avg. length
Lastfm	982	506,582	23,802	7.66
Irvine	540	62,605	471	13.63
Memetracker	498	158,194	8,304	8.43
Twitter	19,546	18,687,423	6,158	36.74

especially for large datasets. We also calculate the average cascade length of each dataset in the last column.

12.3.2   BASELINES

We consider a number of state-of-the-art baselines to demonstrate the effectiveness of our algorithm. Most of the baseline methods will learn a transition probability matrix $M \in \mathbb{R}^{|\mathcal{U}| \times |\mathcal{U}|}$ from cascade sequences where each entry M_{ij} represents the probability that user u_j gets infected by u_i when u_i is activated.

Netrate [Rodriguez et al., 2014] considers the time-varying dynamics of diffusion probability through each link and defines three transmission probability models, i.e., exponential, power-law, and Rayleigh, which encourage the diffusion probability to decrease as the time interval increases. In our experiments, we only report the results of exponential model since the other two models give similar results.

Infopath [Gomez Rodriguez et al., 2013] also targets on inferring dynamic diffusion probabilities based on information diffusion data. Infopath employs stochastic gradient to estimate the temporal dynamics and studies the temporal evolution of information pathways.

Embedded IC [Bourigault et al., 2016] explores representation learning technique and models the diffusion probability between two users by a function of their user embeddings instead of a static value. Embedded IC model is trained by stochastic gradient descent method.

LSTM is a widely used neural network framework [Hochreiter and Schmidhuber, 1997] for sequential data modeling and has been used for cascade modeling recently. Previous works employ LSTM for some simpler tasks such as popularity prediction [Li et al., 2017a] and cascade prediction with known diffusion graph [Hu et al., 2018, Wang et al., 2017c]. Since none of these works are directly comparable to ours, we adopt LSTM network for comparison by adding a softmax classifier to the hidden state of LSTM at each step for next infected user prediction.

12.3.3   HYPERPARAMETER SETTINGS

Though the parameter space of neural network-based methods is much less than that of traditional IC models, we have to set several hyperparameters to train neural models. To tune the hyperparameters, we randomly select 10% of training cascade sequences as validation set. Note that all training cascade sequences including the validation set will be used to train the final model for testing.

For Embedded IC model, the dimension of user embeddings is selected from $\{25, 50, 100\}$ as the original paper did [Bourigault et al., 2016]. For LSTM model, the dimensions of user embeddings and hidden states are set to the best choice from $\{16, 32, 64, 128\}$. For our model NDM, the number of heads used in multi-head attention is set to $h = 8$, the window size of convolutional network is set to $win = 3$ and the dimension of user embeddings is set to $d = 64$. Note that we use the same set of (h, win, d) for all the datasets. The flag F_{init} in Eq. (12.7) which determines whether the initial user is used for prediction is set to $F_{init} = 1$ for Twitter dataset and $F_{init} = 0$ for the other three datasets.

12.3.4   MICROSCOPIC DIFFUSION PREDICTION

To compare the ability of cascade modeling, we evaluate our model and all baseline methods on the microscopic diffusion prediction task. We follow the experimental settings in Embedded IC [Bourigault et al., 2016] for a fair comparison. We randomly select 90% cascade sequences as training set and the rest as test set. For each cascade sequence $c = (u_0, u_1, u_2 \dots)$ in the test set, only the initial user u_0 is given and all successively infected users $G^c = \{u_1, u_2 \dots u_{|G^c|}\}$ need to be predicted. Note that we completely ignore exact timestamp information in this work and the time order among sequences is omitted for simplification. We will explore a more reasonable setting as future work by taking timestamp information into consideration.

All baseline methods and our model are required to predict a set of users and the results will be compared with ground truth infected user set G. For baseline methods that ground in IC model, i.e., Netrate, Infopath, and Embedded IC, we will simulate the infection process according to the learned pairwise diffusion probability and their corresponding generation process. For LSTM and our model, we can sequentially sample a user according to the probability distribution of softmax classifier at each step.

Note that the ground truth infected user set could also be partially observed because the datasets are crawled within a short time window. Therefore, for each test sequence c with $|G^c|$ ground truth infected users, all the algorithms are only required to identify the first $|G^c|$ infected users in a single simulation. Also note that the simulation may terminate and stop infecting new users before activating $|G^c|$ users.

We conduct 1,000 times Monte Carlo simulations for each test cascade sequence c for all algorithms and compute the infection probability P_u^c of each user $u \in \mathcal{U}$. We evaluate the prediction results using two classic evaluation metrics: Macro-F1 and Micro-F1.

Table 12.2: Experimental results on microscopic diffusion prediction

Metric	Dataset	Method					Improvement
		Netrate	Infopath	Embedded IC	LSTM	NDM	
Macro-F1	Lastfm	0.017	0.030	0.020	0.026	**0.056**	+87%
	Memetracker	0.068	0.110	0.060	0.102	**0.139**	+26%
	Irvine	0.032	0.052	0.054	0.041	**0.076**	+41%
	Twitter	—	0.044	—	0.103	**0.139**	+35%
Micro-F1	Lastfm	0.007	0.046	0.085	0.072	**0.095**	+12%
	Memetracker	0.050	0.142	0.115	0.137	**0.171**	+20%
	Irvine	0.029	0.073	0.102	0.080	**0.108**	+6%
	Twitter	—	0.010	—-	0.052	**0.087**	+67%

Table 12.3: Experimental results on microscopic diffusion prediction at early stage where only the first five infected users are predicted in each cascade

Metric	Dataset	Method					Improvement
		Netrate	Infopath	Embedded IC	LSTM	NDM	
Macro-F1	Lastfm	0.018	0.028	0.010	0.018	**0.048**	+71%
	Memetracker	0.071	0.094	0.042	0.091	**0.122**	+30%
	Irvine	0.031	0.030	0.027	0.018	**0.064**	+106%
	Twitter	—	0.040	—	0.097	**0.123**	+27%
Micro-F1	Lastfm	0.016	0.035	0.013	0.019	**0.045**	+29%
	Memetracker	0.076	0.106	0.040	0.094	**0.126**	+19%
	Irvine	0.028	0.030	0.029	0.020	**0.065**	+117%
	Twitter	—	0.050	—	0.093	**0.118**	+27%

To further evaluate the performance of cascade prediction at early stage, we conduct additional experiments by only predicting the first five infected users in each test cascade. We present the experimental results in Tables 12.2 and 12.3. Here "-" indicates that the algorithm fails to converge in 72 h. The last column represents the relative improvement of NDM against the best performing baseline method. We have the following observations.

(1) NDM consistently and significantly outperforms all the baseline methods. As shown in Table 12.2, the relative improvement against the best performing baseline is at least 26%

in terms of Macro-F1 score. The improvement on Micro-F1 score further demonstrates the effectiveness and robustness of our proposed model. The results also indicate that well-designed neural network models are able to surpass traditional cascade methods on cascade modeling.

(2) NDM has even more significant improvements on cascade prediction task *at early stage*. As shown in Table 12.3, NDM outperforms all baselines by a large margin on both Macro and Micro F1 scores. Note that it's very important to identify the first wave of infected users accurately for real-world applications because a wrong prediction will lead to error propagation in following stages. A precise prediction of infected users at early stage enables us to better control the spread of information items through users. For example, we can prevent the spread of a rumor by warning the most vulnerable users in advance and promote the spread of a product by paying the most potential customers more attention. This experiment demonstrates that NDM has the ability to be used for real-world applications.

(3) NDM is capable of handling large-scale cascade datasets. Previous neural-based method, Embedded IC, fails to converge in 72 hours on Twitter dataset with around 20,000 users and 19 million of potential links. In contrast, NDM converges in 6 hours on this dataset with the same GPU device, which is at least 10 times faster than Embedded IC. This observation demonstrates the scalability of NDM.

12.3.5   BENEFITS FROM NETWORK EMBEDDINGS

Sometimes the underlying social network of users is available, e.g., the Twitter dataset used in our experiments. In the Twitter dataset, a network of Twitter followers is observed though the information diffusion is not necessarily passed through the edges of the social network. We hope that the diffusion prediction process could benefit from the observed social network structure. We apply a simple modification on our NDM model to take advantage of the social network. Now we will introduce the modification in detail.

First, we embed the topological social network structure into real-valued user features by DeepWalk [Perozzi et al., 2014], a widely used network representation learning algorithm. The dimension of network embeddings learned by DeepWalk is set to 32 which is half of the dimension $d = 64$ which is the representation size of our model. Second, we use the learned network embeddings to initialize the first 32 dimensions of the user representations of our model and fix them during the training process without changing any other modules. In other words, a 64-dimensional user representation is made up of a 32-dimensional fixed network embedding learned by DeepWalk from social network structure and another 32-dimensional randomly initialized trainable embedding. We name the modified model with social network considered as NDM+SN for short. This is a simple but useful implementation and we will explore a more sophisticated model to take the social network into modeling directly in future work. Figures 12.3 and 12.4 show the comparison between NDM and NDM+SN.

Experimental results show that NDM+SN is able to improve the performance on diffusion prediction task slightly with the help of incorporating social network structure as prior

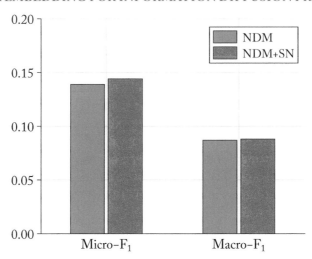

Figure 12.3: Comparisons between NDM and NDM+SN on diffusion prediction.

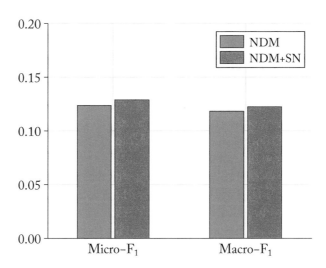

Figure 12.4: Comparisons between NDM and NDM+SN on diffusion prediction at early stage where only the first five infected users are predicted.

knowledge. The relative improvement of Micro-F1 is around 4%. The results demonstrate that our neural model is very flexible and can be easily extended to take advantage of external features.

Table 12.4: The scale of learned projection matrices in convolutional layer measured by Frobenius norm $||\cdot||_F^2$

Dataset	W_{init}^C	W_0^C	W_1^C	W_2^C
Lastfm	32.3	60.0	49.2	49.1
Memetracker	13.3	16.6	13.3	13.0
Irvine	13.9	13.9	13.7	13.7
Twitter	130.3	93.6	91.5	91.5

12.3.6   INTERPRETABILITY

Admittedly, the interpretability is usually a weak point of neural network models. Compared with feature engineering methods, neural-based models encode a user into a real-valued vector space and there is no explicit meaning of each dimension of user embeddings. In our proposed model, each user embedding is projected into 16 subspaces by an 8-head attention mechanism. Intuitively, the user embedding in each subspace represents a specific role of the user. But it is quite hard for us to link the 16 embeddings to interpretable hand-crafted features. We will consider the alignment between user embeddings and interpretable features based on a joint model in future work.

Fortunately, we still have some findings in the convolutional layer. Recall that $W_n^C \in \mathbb{R}^{d \times |\mathcal{U}|}$ for $n = 0, 1, 2$ are position-specific linear projection matrices in convolutional layer and W_{init}^C is the projection matrix for the initial user. All four matrices are randomly initialized before training. In a learned model, if the scale of one of these matrices is much larger than that of other ones, then the prediction vector is more likely to be dominated by the corresponding position. For example, if the scale of W_0^C is much larger than that of other ones, then we can infer that the most recent infected user contributes most to the next infected user prediction.

Following the notations in Eq. (12.7), we set $F_{init} = 1$ for all datasets in this experiment and compute the square of Frobenius norm of learned projection matrices as shown in Table 12.4. We have the following observations.

(1) For all four datasets, the scales of W_0^C, W_1^C, and W_2^C are competitive and the scale of W_0^C is always a little bit larger than that of the other two. This observation indicates that the active embeddings $act(u_j), act(u_{j-1}), act(u_{j-2})$ of all three recently infected users will contribute to the prediction of u_{j+1}. Also, the most recent infected user u_j is the most important one among the three. This finding naturally matches our intuitions and verifies *Assumption 2* proposed in method section.

(2) The scale of W_{init}^C is the largest on Twitter dataset. This indicates that the initial user is very important in diffusion process on Twitter. This is partly because Twitter dataset contains the complete history of the spread of a URL and the initial user is actually the first one tweeting the URL. While in the other three datasets, the initial user is only the first one within the time

window of crawled data. Note that we set hyperparameter $F_{init} = 1$ only for Twitter dataset in diffusion prediction task because we find that the performances are competitive or even worse on the other three datasets if we set $F_{init} = 1$.

12.4 FURTHER READING

Information diffusion prediction methods can be categorized into macroscopic and microscopic cascade studies.

Macroscopic Cascade Study: Most previous works focused on macroscopic level estimation such as the eventual size of a cascade [Zhao et al., 2015b] and the growth curve of popularity [Yu et al., 2015]. Macroscopic cascade study methods can be further classified into feature-based approaches, generative approaches, and deep-learning based approaches. Feature-based approaches formalized the task as a classification problem [Cheng et al., 2014, Cui et al., 2013] or a regression problem [Tsur and Rappoport, 2012, Weng et al., 2014] by applying SVM, logistic regression, and other machine learning algorithms on hand-crafted features including temporal [Pinto et al., 2013] and structural [Cheng et al., 2014] features. Generative approaches considered the growth of cascade size as an arrival process of infected users and employed stochastic processes, such as Hawkes self-exciting point process [Gao et al., 2015b, Zhao et al., 2015b], for modeling. With the success of deep learning techniques in various applications, deep-learning based approaches, e.g., DeepCas [Li et al., 2017a] and DeepHawkes [Cao et al., 2017], were proposed to employ RNN for encoding cascade sequences into feature vectors instead of hand-crafted features. Compared with hand-crafted feature engineering, deep-learning based approaches have better generalization ability across different platforms and give better performance on evaluation tasks.

Microscopic Cascade Study: Microscopic cascade study focuses on user-level modeling and can be further classified into three groups: IC-based approaches, embedding-based approaches, and deep-learning based approaches.

IC model [Goldenberg et al., 2001, Gruhl et al., 2004, Kempe et al., 2003, Saito et al., 2008] is one of the most popular diffusion models which assumed independent diffusion probability through each link. Extensions of IC model further considered time delay information by incorporating a predefined time-decay weighting function, such as continuous time IC [Saito et al., 2009], CONNIE [Myers and Leskovec, 2010], NetInf [Gomez-Rodriguez et al., 2012], and Netrate [Rodriguez et al., 2014]. Infopath [Gomez Rodriguez et al., 2013] was proposed to infer dynamic diffusion probabilities based on information diffusion data and study the temporal evolution of information pathways. MMRate [Wang et al., 2014] inferred multi-aspect transmission rates by incorporating aspect-level user interactions and various diffusion patterns. All of the above methods learned the probabilities from cascade sequences. Once a model is trained, it can be used for microscopic evaluation tasks by simulating the generative process using Monte Carlo simulation.

Embedding-based approaches encoded each user into a parameterized real-valued vector and trained the parameters by maximizing an objective function. Embedded IC [Bourigault et al., 2016] followed the pairwise independence assumption in IC model and modeled the diffusion probability between two users by a function of their user embeddings. Other embedding-based diffusion models [Bourigault et al., 2014, Gao et al., 2017b] made even stronger assumptions that infected users are determined only by the source user and the content of information item. However, embedding-based methods failed to model the infection history, e.g., the ordering of infected users, for next infected user prediction and have been shown to be suboptimal choices in the experiments of recent deep-learning based approaches.

For deep-learning based approaches, [Hu et al., 2018] employed LSTM for modeling and focused on the retweeting behaviors in a social network where "who infected whom" information is explicitly labeled in retweet chains. However, the diffusion graph is usually unknown for most diffusion processes [Myers and Leskovec, 2010, Wallinga and Teunis, 2004]. TopoL-STM [Wang et al., 2017c] extended the standard LSTM model by structuring the hidden states as a directed acyclic graph extracted from the social graph. CYAN-RNN [Wang et al., 2017h], DAN [Wang et al., 2018b], and DeepDiffuse [Islam et al., 2018] all employed RNN and attention model to utilize the infection timestamp information. SNIDSA [Wang et al., 2018c] computed pairwise similarities of all user pairs and incorporated the structural information into RNN by a gating mechanism. During the last five years, deep-learning based approaches have become the main stream in this field.

Part of this chapter was from our arxiv paper by Yang et al. [2018a].

PART V

Outlook for Network Embedding

CHAPTER 13

Future Directions of Network Embedding

The aforementioned network embedding methods have demonstrated their effectiveness in various scenarios and applications. With the rapid growth of data scales and the development of deep learning techniques, there are also new challenges and opportunities for next-stage researches of network embedding. In the last chapter, we will look into the future directions of NRL. Specifically, we will consider the following directions including employing advanced techniques, considering fine-grained scenarios and utilizing network embedding for specific applications.

13.1 NETWORK EMBEDDING BASED ON ADVANCED TECHNIQUES

As deep learning techniques get popular and achieve promising results in many areas, we have witnessed the rising of a number of deep learning models such as variational autoencoder (VAE) [Kingma and Welling, 2013] and generative adversarial networks (GAN) [Goodfellow et al., 2014]. These advanced models are adopted for network representation learning in follow-up works, e.g., GraphVAE [Simonovsky and Komodakis, 2018], GraphGAN [Wang et al., 2017b], and GraphSGAN [Ding et al., 2018]. Recently, there is also a trend of incorporating contrastive learning technique [Hafidi et al., 2020, Hassani and Khasahmadi, 2020, Zhu et al., 2020] into network embedding. On the other direction, GNN technique in graph learning area was widely employed in computer vision and natural language processing areas. Therefore, it would never be out-of-date to employ the most advanced techniques for modeling.

13.2 NETWORK EMBEDDING IN MORE FINE-GRAINED SCENARIOS

The study of network embedding methods has a long history of almost 20 years. As the main branch and the most general case of network embedding, learning representations with only the topology structure of a network has been well studied and reached a bottleneck limited by current deep learning techniques and computing capabilities.

However, there is still some room for improvement on more fine-grained scenarios. In fact, most of this book focuses on these fine-grained settings, such as attributed network embedding and network embedding specialized for large-scale graphs. A real-world network is usually highly dynamic, heterogeneous, and large-scale. Though existing works have considered these characteristics separately, a combination of these properties would raise many new challenges for network embedding and a trivial combinative modeling may not work well in practice. For example, network representation learning on dynamic large-scale graphs should deal with the embedding learning of new incoming nodes without retraining the whole model (also known as incremental learning). Thus, an important future direction is to develop network embeddings in a more realistic and fine-grained scenario, which makes learned representations more applicable to real-world services.

13.3   NETWORK EMBEDDING WITH BETTER INTERPRETABILITY

An important research topic in artificial intelligence is model transparency and interpretability. For decision-critical applications related to ethics, privacy, and safety, it is necessary for models to generate human interpretable predictions. However, different from hand-crafted node features such as node degree and betweenness coefficient, most learned network representations are short of model transparency, especially for the deep models such as SDNE [Wang et al., 2016a]. In other words, we have no idea how each dimension of network embeddding links to the raw inputs. In order to utilize network embedding techniques for decision-critical applications, there is a need to improve model interpretability and transparency of current methods. Recently, GN-NExplainer [Ying et al., 2019] were proposed to interpret the predictions of GNN and we are looking forward to more studies on improving model interpretability to facilitate the extensive use of network embeddings in various domains.

13.4   NETWORK EMBEDDING FOR APPLICATIONS

A practical research direction is to utilize network embedding techniques for specific applications rather than proposing a new network representation model. Note that both sides are equally important for researches on network analysis and the industrial applications will also inspire the development of NRL methods in return. We present several applications of network embedding in Part IV of this book, including social relation extraction, recommendation system, and information diffusion prediction.

In fact, the applications of network embedding should not be restricted in machine learning and data mining areas. Graph-structured data are ubiquitous in various domains, such as the modeling of molecules or proteins in biochemistry. For instance, there has been some recent work [You et al., 2018] which targets on graph generation to discover novel molecules with desired properties. These cross-domain applications of network embedding will make a strong impact on relevant researches.

Bibliography

Y.-Y. Ahn, J. P. Bagrow, and S. Lehmann. 2010. Link communities reveal multiscale complexity in networks. *Nature*. DOI: 10.1038/nature09182 94, 98

E. M. Airoldi, D. M. Blei, S. E. Fienberg, and E. P. Xing. 2008. Mixed membership stochastic blockmodels. *JMLR*, 9(Sep):1981–2014. 67

E. Akbas and M. E. Aktas. 2019. Network embedding: On compression and learning. In *IEEE International Conference on Big Data (Big Data)*, pages 4763–4772. DOI: 10.1109/bigdata47090.2019.9006142 116

R. Andersen, F. Chung, and K. Lang. 2006. Local graph partitioning using pagerank vectors. In *Proc. of FOCS*, pages 475–486, IEEE. DOI: 10.1109/focs.2006.44 10

S. Aral and D. Walker. 2012. Identifying influential and susceptible members of social networks. *Science*. DOI: 10.1126/science.1215842 171

S. Auer, C. Bizer, G. Kobilarov, J. Lehmann, R. Cyganiak, and Z. Ives. 2007. Dbpedia: A nucleus for a Web of open data. In *The Semantic Web*, pages 722–735. DOI: 10.1007/978-3-540-76298-0_52 145

D. Bahdanau, K. Cho, and Y. Bengio. 2015. Neural machine translation by jointly learning to align and translate. In *Proc. of ICLR*. 176

M. Balcilar, G. Renton, P. Héroux, B. Gauzere, S. Adam, and P. Honeine. 2020. Bridging the gap between spectral and spatial domains in graph neural networks. *ArXiv Preprint ArXiv:2003.11702*. 57

J. Bao, Y. Zheng, and M. F. Mokbel. 2012. Location-based and preference-aware recommendation using sparse geo-social networking data. In *Proc. of the 20th International Conference on Advances in Geographic Information Systems*, pages 199–208, ACM. DOI: 10.1145/2424321.2424348 148

J. Bao, Y. Zheng, D. Wilkie, and M. Mokbel. 2015. Recommendations in location-based social networks: A survey. *GeoInformatica*, 19(3):525–565. DOI: 10.1007/s10707-014-0220-8 168

M. Belkin and P. Niyogi. 2001. Laplacian eigenmaps and spectral techniques for embedding and clustering. In *Proc. of NeurIPS*, 14:585–591. DOI: 10.7551/mitpress/1120.003.0080 5, 9

A. Ben-Hur and J. Weston. 2010. A user's guide to support vector machines. *Data Mining Techniques for the Life Sciences*. DOI: 10.1007/978-1-60327-241-4_13 22

Y. Bengio, J. Louradour, R. Collobert, and J. Weston. 2009. Curriculum learning. In *Proc. of ICML*, pages 41–48. DOI: 10.1145/1553374.1553380 47

Y. Bengio, A. Courville, and P. Vincent. 2013. Representation learning: A review and new perspectives. *IEEE Transactions on Pattern Analysis and Machine Intelligence*, 35(8):1798–1828. DOI: 10.1109/tpami.2013.50 5

P. Bhargava, T. Phan, J. Zhou, and J. Lee. 2015. Who, what, when, and where: Multi-dimensional collaborative recommendations using tensor factorization on sparse user-generated data. In *Proc. of the 24th International Conference on World Wide Web*, pages 130–140, ACM. DOI: 10.1145/2736277.2741077 169

M. H. Bhuyan, D. Bhattacharyya, and J. K. Kalita. 2014. Network anomaly detection: Methods, systems and tools. *IEEE Communications Surveys and Tutorials*, 16(1):303–336. DOI: 10.1109/surv.2013.052213.00046 3

D. M. Blei, A. Y. Ng, and M. I. Jordan. 2003. Latent Dirichlet allocation. *JMLR*, 3:993–1022. DOI: 10.1109/asru.2015.7404785 83

V. D. Blondel, J.-L. Guillaume, R. Lambiotte, and E. Lefebvre. 2008. Fast unfolding of communities in large networks. *JSTAT*. DOI: 10.1088/1742-5468/2008/10/p10008 97

P. Blunsom, E. Grefenstette, and N. Kalchbrenner. 2014. A convolutional neural network for modelling sentences. In *Proc. of ACL*. DOI: 10.3115/v1/p14-1062 60, 63

K. Bollacker, C. Evans, P. Paritosh, T. Sturge, and J. Taylor. 2008. Freebase: A collaboratively created graph database for structuring human knowledge. In *Proc. of SIGMOD*, pages 1247–1250. DOI: 10.1145/1376616.1376746 145

A. Bordes, N. Usunier, A. Garcia-Duran, J. Weston, and O. Yakhnenko. 2013. Translating embeddings for modeling multi-relational data. In *Proc. of NIPS*, pages 2787–2795. 125, 126, 129, 135, 136, 137, 141, 142, 143, 145

J. A. Botha, E. Pitler, J. Ma, A. Bakalov, A. Salcianu, D. Weiss, R. McDonald, and S. Petrov. 2017. Natural language processing with small feed-forward networks. *ArXiv Preprint ArXiv:1708.00214*. DOI: 10.18653/v1/d17-1309 117

S. Bourigault, C. Lagnier, S. Lamprier, L. Denoyer, and P. Gallinari. 2014. Learning social network embeddings for predicting information diffusion. In *Proc. of WSDM*, ACM. DOI: 10.1145/2556195.2556216 172, 187

S. Bourigault, S. Lamprier, and P. Gallinari. 2016. Representation learning for information diffusion through social networks: An embedded cascade model. In *Proc. of WSDM*, ACM. DOI: 10.1145/2835776.2835817 172, 173, 174, 178, 179, 180, 181, 187

A. Broder, R. Kumar, F. Maghoul, P. Raghavan, S. Rajagopalan, R. Stata, A. Tomkins, and J. Wiener. 2000. Graph structure in the Web. *Computer Networks*, 33(1–6): 309–320. DOI: 10.1016/s1389-1286(00)00083-9 100

Q. Cao, H. Shen, K. Cen, W. Ouyang, and X. Cheng. 2017. Deephawkes: Bridging the gap between prediction and understanding of information cascades. In *Proc. of CIKM*, ACM. DOI: 10.1145/3132847.3132973 186

S. Cao, W. Lu, and Q. Xu. 2015. GraRep: Learning graph representations with global structural information. In *Proc. of CIKM*. DOI: 10.1145/2806416.2806512 13, 14, 16, 22

S. Cavallari, V. W. Zheng, H. Cai, K. C.-C. Chang, and E. Cambria. 2017. Learning community embedding with community detection and node embedding on graphs. In *Proc. of the ACM on Conference on Information and Knowledge Management*, pages 377–386. DOI: 10.1145/3132847.3132925 88, 98

O. Celma. 2010. *Music Recommendation and Discovery in the Long Tail*. Springer. DOI: 10.1007/978-3-642-13287-2 179

J. Chang, L. Wang, G. Meng, S. Xiang, and C. Pan. 2017. Deep adaptive image clustering. In *Proc. of ICCV*, pages 5879–5887. DOI: 10.1109/iccv.2017.626 42, 46

S. Chang, W. Han, J. Tang, G.-J. Qi, C. C. Aggarwal, and T. S. Huang. 2015. Heterogeneous network embedding via deep architectures. In *Proc. of SIGKDD*, pages 119–128, ACM. DOI: 10.1145/2783258.2783296 120, 131, 132

H. Chen, B. Perozzi, Y. Hu, and S. Skiena. 2017. Harp: Hierarchical representation learning for networks. *ArXiv Preprint ArXiv:1706.07845*. 103, 116

H. Chen, H. Yin, W. Wang, H. Wang, Q. V. H. Nguyen, and X. Li. 2018. PME: Projected metric embedding on heterogeneous networks for link prediction. In *Proceeding of SIGKDD*, pages 1177–1186. DOI: 10.1145/3219819.3219986 131, 132

M. Chen, Q. Yang, and X. Tang. 2007. Directed graph embedding. In *Proc. of IJCAI*, pages 2707–2712. DOI: 10.1101/110668 5, 10, 37, 145

W. Chen, J. Wilson, S. Tyree, K. Weinberger, and Y. Chen. 2015. Compressing neural networks with the hashing trick. In *International Conference on Machine Learning*, pages 2285–2294. 117

X. Chen, Y. Duan, R. Houthooft, J. Schulman, I. Sutskever, and P. Abbeel. 2016. Infogan: Interpretable representation learning by information maximizing generative adversarial nets. In *Advances in Neural Information Processing Systems*, pages 2172–2180. 73

Z. Chen, F. Chen, L. Zhang, T. Ji, K. Fu, L. Zhao, F. Chen, and C.-T. Lu. 2020. Bridging the gap between spatial and spectral domains: A survey on graph neural networks. *ArXiv Preprint ArXiv:2002.11867*. 57

C. Cheng, H. Yang, I. King, and M. R. Lyu. 2012. Fused matrix factorization with geographical and social influence in location-based social networks. In *AAAI*, 12:17–23. 152, 164, 168

C. Cheng, H. Yang, M. R. Lyu, and I. King. 2013. Where you like to go next: Successive point-of-interest recommendation. In *Proc. of IJCAI*. 150, 160, 161, 169

J. Cheng, L. Adamic, P. A. Dow, J. M. Kleinberg, and J. Leskovec. 2014. Can cascades be predicted? In *Proc. of WWW*, ACM. DOI: 10.1145/2566486.2567997 171, 172, 186

E. Cho, S. A. Myers, and J. Leskovec. 2011. Friendship and mobility: User movement in location-based social networks. In *Proc. of SIGKDD*. DOI: 10.1145/2020408.2020579 148, 160

Y.-S. Cho, G. Ver Steeg, and A. Galstyan. 2013. Socially relevant venue clustering from check-in data. In *KDD Workshop on Mining and Learning with Graphs*. 148

F. R. Chung and F. C. Graham. 1997. *Spectral Graph Theory*. American Mathematical Society. DOI: 10.1090/cbms/092 46

R. Collobert and J. Weston. 2008. A unified architecture for natural language processing: Deep neural networks with multitask learning. In *Proc. of ICML*, ACM. DOI: 10.1145/1390156.1390177 177

K. Crammer and Y. Singer. 2002. On the learnability and design of output codes for multiclass problems. *Machine Learning*, 47(2–3):201–233. DOI: 10.1023/A:1013637720281 80

G. Cui, J. Zhou, C. Yang, and Z. Liu. 2020. Adaptive graph encoder for attributed graph embedding. In *Proc. of SIGKDD*. DOI: 10.1145/3394486.3403140 xxi, 57

P. Cui, S. Jin, L. Yu, F. Wang, W. Zhu, and S. Yang. 2013. Cascading outbreak prediction in networks: A data-driven approach. In *Proc. of SIGKDD*, ACM. DOI: 10.1145/2487575.2487639 186

A. Dalmia, M. Gupta, et al. 2018. Towards interpretation of node embeddings. In *Companion of the Web Conference*, pages 945–952, International World Wide Web Conferences Steering Committee. DOI: 10.1145/3184558.3191523 110

P.-E. Danielsson. 1980. Euclidean distance mapping. *Computer Graphics and Image Processing*, 14(3):227–248. DOI: 10.1016/0146-664x(80)90054-4 124

D. L. Davies and D. W. Bouldin. 1979. A cluster separation measure. *IEEE Transactions on Pattern Analysis and Machine Intelligence*, (2):224–227. DOI: 10.1109/tpami.1979.4766909 7, 48

C. Deng, Z. Zhao, Y. Wang, Z. Zhang, and Z. Feng. 2019. GraphZoom: A multi-level spectral approach for accurate and scalable graph embedding. *ArXiv Preprint ArXiv:1910.02370.* 116

M. Denil, B. Shakibi, L. Dinh, N. De Freitas, et al. 2013. Predicting parameters in deep learning. In *Advances in Neural Information Processing Systems*, pages 2148–2156. 116

M. Ding, J. Tang, and J. Zhang. 2018. Semi-supervised learning on graphs with generative adversarial nets. In *Proc. of the 27th ACM International Conference on Information and Knowledge Management*, pages 913–922. DOI: 10.1145/3269206.3271768 191

P. Domingos and M. Richardson. 2001. Mining the network value of customers. In *Proc. of SIGKDD*, ACM. DOI: 10.1145/502512.502525 171

Y. Dong, N. V. Chawla, and A. Swami. 2017. metapath2vec: Scalable representation learning for heterogeneous networks. In *Proc. of SIGKDD*, pages 135–144, ACM. DOI: 10.1145/3097983.3098036 119, 122, 127, 129, 131

C. N. dos Santos, M. Tan, B. Xiang, and B. Zhou. 2016. Attentive pooling networks. *CoRR, abs/1602.03609.* 63, 72

P. A. Dow, L. A. Adamic, and A. Friggeri. 2013. The anatomy of large Facebook cascades. In *8th International AAAI Conference on Weblogs and Social Media*, pages 145–154. 172

L. Du, Z. Lu, Y. Wang, G. Song, Y. Wang, and W. Chen. 2018. Galaxy network embedding: A hierarchical community structure preserving approach. In *IJCAI*, pages 2079–2085. DOI: 10.24963/ijcai.2018/287 98

J. Duchi, E. Hazan, and Y. Singer. 2011. Adaptive subgradient methods for online learning and stochastic optimization. *The Journal of Machine Learning Research*, 12:2121–2159. 159

A. Epasto and B. Perozzi. 2019. Is a single embedding enough? Learning node representations that capture multiple social contexts. In *The World Wide Web Conference*, pages 394–404. DOI: 10.1145/3308558.3313660 72, 73

R.-E. Fan, K.-W. Chang, C.-J. Hsieh, X.-R. Wang, and C.-J. Lin. 2008. Liblinear: A library for large linear classification. *JMLR.* 22, 34, 67, 94, 109

W. Fan, Y. Ma, Q. Li, Y. He, E. Zhao, J. Tang, and D. Yin. 2019. Graph neural networks for social recommendation. In *The World Wide Web Conference*, pages 417–426. DOI: 10.1145/3308558.3313488 170

Y. Fan, S. Hou, Y. Zhang, Y. Ye, and M. Abdulhayoglu. 2018. Gotcha—sly malware!: Scorpion a metagraph2vec based malware detection system. In *Proc. of SIGKDD*, pages 253–262. DOI: 10.1145/3219819.3219862 132

K. Faust. 1997. Centrality in affiliation networks. *Social Networks*, 19(2):157–191. DOI: 10.1016/s0378-8733(96)00300-0 123, 124, 125

S. Feng, X. Li, Y. Zeng, G. Cong, Y. M. Chee, and Q. Yuan. 2015. Personalized ranking metric embedding for next new poi recommendation. In *Proc. of IJCAI*. 160, 161, 162, 165, 167, 169

S. Fortunato. 2010. Community detection in graphs. *Physics Reports*. DOI: 0.1016/j.physrep.2009.11.002 98

F. Fouss, A. Pirotte, J.-M. Renders, and M. Saerens. 2007. Random-walk computation of similarities between nodes of a graph with application to collaborative recommendation. *IEEE Transactions on Knowledge and Data Engineering*, 19(3):355–369. DOI: 10.1109/tkde.2007.46 10

T.-Y. Fu, W.-C. Lee, and Z. Lei. 2017. Hin2vec: Explore meta-paths in heterogeneous information networks for representation learning. In *Proc. of CIKM*, pages 1797–1806, ACM. DOI: 10.1145/3132847.3132953 119, 127, 131

G. Gan, C. Ma, and J. Wu. 2007. *Data Clustering: Theory, Algorithms, and Applications*, vol. 20. SIAM. DOI: 10.1137/1.9780898718348 7, 50

H. Gao and H. Huang. 2018. Deep attributed network embedding. In *IJCAI*, 18:3364–3370, New York. DOI: 10.24963/ijcai.2018/467 38

H. Gao, J. Tang, X. Hu, and H. Liu. 2013. Exploring temporal effects for location recommendation on location-based social networks. In *Proc. of the 7th ACM Conference on Recommender Systems*, pages 93–100, ACM. DOI: 10.1145/2507157.2507182 169

H. Gao, J. Tang, X. Hu, and H. Liu. 2015a. Content-aware point of interest recommendation on location-based social networks. In *AAAI*, pages 1721–1727, CiteSeer. 164, 169

Q. Gao, F. Zhou, K. Zhang, G. Trajcevski, X. Luo, and F. Zhang. 2017a. Identifying human mobility via trajectory embeddings. In *IJCAI*, 17:1689–1695. DOI: 10.24963/ijcai.2017/234 169

S. Gao, J. Ma, and Z. Chen. 2015b. Modeling and predicting retweeting dynamics on microblogging platforms. In *Proc. of WSDM*, ACM. DOI: 10.1145/2684822.2685303 186

S. Gao, H. Pang, P. Gallinari, J. Guo, and N. Kato. 2017b. A novel embedding method for information diffusion prediction in social network big data. *IEEE Transactions on Industrial Informatics*. DOI: 10.1109/tii.2017.2684160 173, 187

J. Goldenberg, B. Libai, and E. Muller. 2001. Talk of the network: A complex systems look at the underlying process of word-of-mouth. *Marketing Letters*. DOI: 10.1023/A:1011122126881 186

M. Gomez-Rodriguez, J. Leskovec, and A. Krause. 2012. Inferring networks of diffusion and influence. *ACM Transactions on Knowledge Discovery from Data (TKDD)*, 5(4):21. DOI: 10.1145/2086737.2086741 172, 186

M. Gomez Rodriguez, J. Leskovec, and B. Schölkopf. 2013. Structure and dynamics of information pathways in online media. In *Proc. of WSDM*, ACM. DOI: 10.1145/2433396.2433402 172, 180, 186

I. Goodfellow, J. Pouget-Abadie, M. Mirza, B. Xu, D. Warde-Farley, S. Ozair, A. Courville, and Y. Bengio. 2014. Generative adversarial nets. In *Advances in Neural Information Processing Systems*, pages 2672–2680. DOI: 10.1145/3422622 191

T. L. Griffiths and M. Steyvers. 2004. Finding scientific topics. *PNAS*. DOI: 10.1073/pnas.0307752101 91

A. Grover and J. Leskovec. 2016. node2vec: Scalable feature learning for networks. DOI: 10.1145/2939672.2939754 3, 5, 12, 17, 21, 22, 25, 60, 67, 88, 101, 108, 141

D. Gruhl, R. Guha, D. Liben-Nowell, and A. Tomkins. 2004. Information diffusion through blogspace. In *Proc. of WWW*. DOI: 10.1145/988672.988739 172, 186

H. Hafidi, M. Ghogho, P. Ciblat, and A. Swami. 2020. Graphcl: Contrastive self-supervised learning of graph representations. *ArXiv Preprint ArXiv:2007.08025*. 191

W. Hamilton, Z. Ying, and J. Leskovec. 2017a. Inductive representation learning on large graphs. In *Proc. of NIPS*, pages 1024–1034. 105

W. L. Hamilton, R. Ying, and J. Leskovec. 2017b. Representation learning on graphs: Methods and applications. *IEEE Data(base) Engineering Bulletin*, 40(3):52–74. 99, 100

S. Han, H. Mao, and W. J. Dally. 2015a. Deep compression: Compressing deep neural networks with pruning, trained quantization and huffman coding. *ArXiv Preprint ArXiv:1510.00149*. 117

S. Han, J. Pool, J. Tran, and W. Dally. 2015b. Learning both weights and connections for efficient neural network. In *Advances in Neural Information Processing Systems*, pages 1135–1143. 117

X. Han, C. Shi, S. Wang, S. Y. Philip, and L. Song. 2018. Aspect-level deep collaborative filtering via heterogeneous information networks. In *Proc. of IJCAI*, pages 3393–3399. DOI: 10.24963/ijcai.2018/471 120, 131

J. A. Hanley and B. J. McNeil. 1982. The meaning and use of the area under a receiver operating characteristic (roc) curve. *Radiology*. DOI: 10.1148/radiology.143.1.7063747 7, 23, 94, 109

K. Hassani and A. H. Khasahmadi. 2020. Contrastive multi-view representation learning on graphs. *ArXiv Preprint ArXiv:2006.05582*. 191

M. A. Hearst, S. T. Dumais, E. Osman, J. Platt, and B. Scholkopf. 1998. Support vector machines. *IEEE Intelligent Systems and their Applications*, 13(4):18–28. DOI: 10.1109/5254.708428 75, 77

H. W. Hethcote. 2000. The mathematics of infectious diseases. *SIAM Review*, 42(4):599–653. DOI: 10.1137/s0036144500371907 171

S. Hochreiter and J. Schmidhuber. 1997. Long short-term memory. *Neural Computation*, 9(8):1735–1780. DOI: 10.1162/neco.1997.9.8.1735 180

N. O. Hodas and K. Lerman. 2014. The simple rules of social contagion. *Scientific Reports*, 4:4343. DOI: 10.1038/srep04343 179

R. Hoffmann, C. Zhang, X. Ling, L. Zettlemoyer, and D. S. Weld. 2011. Knowledge-based weak supervision for information extraction of overlapping relations. In *Proc. of ACL-HLT*, pages 541–550. 145

T. Hofmann. 1999. Probabilistic latent semantic indexing. In *Proc. of the 22nd Annual International ACM SIGIR Conference on Research and Development in Information Retrieval*, pages 50–57. DOI: 10.1145/312624.312649 34

R. A. Horn and C. R. Johnson. 2012. *Matrix Analysis*. Cambridge University Press. DOI: 10.1017/cbo9780511810817 43

C.-K. Hsieh, L. Yang, Y. Cui, T.-Y. Lin, S. Belongie, and D. Estrin. 2017. Collaborative metric learning. In *Proc. of WWW*, pages 193–201. DOI: 10.1145/3038912.3052639 124

W. Hu, K. K. Singh, F. Xiao, J. Han, C.-N. Chuah, and Y. J. Lee. 2018. Who will share my image?: Predicting the content diffusion path in online social networks. In *Proc. of the 11th ACM International Conference on Web Search and Data Mining*, pages 252–260. DOI: 10.1145/3159652.3159705 173, 180, 187

H. Huang, J. Tang, S. Wu, L. Liu, et al. 2014. Mining triadic closure patterns in social networks. In *Proc. of the 23rd International Conference on World Wide Web*, pages 499–504, ACM. DOI: 10.1145/2567948.2576940

X. Huang, J. Li, N. Zou, and X. Hu. 2018. A general embedding framework for heterogeneous information learning in large-scale networks. *ACM Transactions on Knowledge Discovery from Data (TKDD)*, 12(6):1–24. DOI: 10.1145/3241063 25

I. Hubara, M. Courbariaux, D. Soudry, R. El-Yaniv, and Y. Bengio. 2016. Quantized neural networks: Training neural networks with low precision weights and activations. *ArXiv Preprint ArXiv:1609.07061.* 117

M. R. Islam, S. Muthiah, B. Adhikari, B. A. Prakash, and N. Ramakrishnan. 2018. DeepDiffuse: Predicting the "who" and "when" in cascades. In *Proc. of ICDM*. DOI: 10.1109/icdm.2018.00134 187

Y. Jacob, L. Denoyer, and P. Gallinari. 2014. Learning latent representations of nodes for classifying in heterogeneous social networks. In *Proc. of WSDM*, pages 373–382, ACM. DOI: 10.1145/2556195.2556225 131

M. Jaderberg, A. Vedaldi, and A. Zisserman. 2014. Speeding up convolutional neural networks with low rank expansions. *ArXiv Preprint ArXiv:1405.3866.* DOI: 10.5244/c.28.88 116

T. Joachims. 1999. Making large-scale support vector machine learning practical. In *Advances in Kernel Methods: Support Vector Learning*, pages 169–184. 34

R. Johnson and T. Zhang. 2014. Effective use of word order for text categorization with convolutional neural networks. *ArXiv Preprint ArXiv:1412.1058.* DOI: 10.3115/v1/n15-1011 60, 63

S. S. Keerthi, S. Sundararajan, K.-W. Chang, C.-J. Hsieh, and C.-J. Lin. 2008. A sequential dual method for large scale multi-class linear SVMs. In *Proc. of the 14th ACM SIGKDD International Conference on Knowledge Discovery and Data Mining*, pages 408–416. DOI: 10.1145/1401890.1401942 79

D. Kempe, J. Kleinberg, and É. Tardos. 2003. Maximizing the spread of influence through a social network. In *Proc. of SIGKDD*, pages 137–146, ACM. DOI: 10.1145/956750.956769 172, 174, 186

B. W. Kernighan and S. Lin. 1970. An efficient heuristic procedure for partitioning graphs. *The Bell System Technical Journal*, 49(2):291–307. DOI: 10.1002/j.1538-7305.1970.tb01770.x 98

Y. Kim. 2014. Convolutional neural networks for sentence classification. DOI: 10.3115/v1/d14-1181 60, 63

Y. Kim, Y. Jernite, D. Sontag, and A. M. Rush. 2016. Character-aware neural language models. In *AAAI*, pages 2741–2749. 117

D. Kingma and J. Ba. 2015. Adam: A method for stochastic optimization. In *Proc. of ICLR*. 50, 65, 140

D. P. Kingma and M. Welling. 2013. Auto-encoding variational Bayes. *ArXiv Preprint ArXiv:1312.6114*. 191

T. N. Kipf and M. Welling. 2016. Variational graph auto-encoders. In *NIPS Workshop on Bayesian Deep Learning*. 41, 42, 49

T. N. Kipf and M. Welling. 2017. Semi-supervised classification with graph convolutional networks. In *Proc. of ICLR*. 39, 40, 83, 101, 105, 169

R. Kiros, Y. Zhu, R. R. Salakhutdinov, R. Zemel, R. Urtasun, A. Torralba, and S. Fidler. 2015. Skip-thought vectors. In *Proc. of NIPS*, pages 3294–3302. 60, 63

H. W. Kuhn. 1955. The Hungarian method for the assignment problem. *Naval Research Logistics Quarterly*, 2(1–2):83–97. DOI: 10.1002/nav.3800020109 7

J. M. Kumpula, M. Kivelä, K. Kaski, and J. Saramäki. 2008. Sequential algorithm for fast clique percolation. *Physical Review E*. DOI: 10.1103/physreve.78.026109 94

Y.-A. Lai, C.-C. Hsu, W. H. Chen, M.-Y. Yeh, and S.-D. Lin. 2017. Prune: Preserving proximity and global ranking for network embedding. In *Advances in Neural Information Processing Systems*, pages 5257–5266. 98

M. Lam. 2018. Word2bits-quantized word vectors. *ArXiv Preprint ArXiv:1803.05651*. 117

T. Lappas, E. Terzi, D. Gunopulos, and H. Mannila. 2010. Finding effectors in social networks. In *Proc. of SIGKDD*. DOI: 10.1145/1835804.1835937 172

D. LaSalle and G. Karypis. 2013. Multi-threaded graph partitioning. In *Parallel and Distributed Processing (IPDPS), IEEE 27th International Symposium on*, pages 225–236. DOI: 10.1109/ipdps.2013.50 115

Q. V. Le and T. Mikolov. 2014. Distributed representations of sentences and documents. *Computer Science*, 4:1188–1196. 161

Y. LeCun et al. 2015. Lenet-5, convolutional neural networks. http://yann.lecun.com/exdb/lenet 177

J. Leskovec, J. Kleinberg, and C. Faloutsos. 2005. Graphs over time: Densification laws, shrinking diameters and possible explanations. In *Proc. of KDD*, pages 177–187. DOI: 10.1145/1081870.1081893 66

J. Leskovec, A. Singh, and J. Kleinberg. 2006. Patterns of influence in a recommendation network. In *Pacific-Asia Conference on Knowledge Discovery and Data Mining*, pages 380–389, Springer. DOI: 10.1007/11731139_44 171

J. Leskovec, L. A. Adamic, and B. A. Huberman. 2007. The dynamics of viral marketing. *ACM Transactions on the Web (TWEB)*, 1(1):5. DOI: 10.1145/1232722.1232727 171

J. Leskovec, L. Backstrom, and J. Kleinberg. 2009. Meme-tracking and the dynamics of the news cycle. In *Proc. of SIGKDD*. DOI: 10.1145/1557019.1557077 172

J. J. Levandoski, M. Sarwat, A. Eldawy, and M. F. Mokbel. 2012. Lars: A location-aware recommender system. In *IEEE 28th International Conference on Data Engineering*, pages 450–461. DOI: 10.1109/icde.2012.54 152, 168

O. Levy and Y. Goldberg. 2014. Neural word embedding as implicit matrix factorization. In *Proc. of NIPS*. 15

M. Ley. 2002. The DBLP computer science bibliography: Evolution, research issues, perspectives. In *International Symposium on String Processing and Information Retrieval*. DOI: 10.1007/3-540-45735-6_1 3

C. Li, J. Ma, X. Guo, and Q. Mei. 2017a. DeepCas: An end-to-end predictor of information cascades. In *Proc. of WWW*. DOI: 10.1145/3038912.3052643 173, 180, 186

G. Li, Q. Chen, B. Zheng, H. Yin, Q. V. H. Nguyen, and X. Zhou. 2020. Group-based recurrent neural networks for poi recommendation. *ACM Transactions on Data Science*, 1(1):1–18. DOI: 10.1145/3343037 169

J. Li, J. Zhu, and B. Zhang. 2016. Discriminative deep random walk for network classification. In *Proc. of ACL*. DOI: 10.18653/v1/p16-1095 83

J. Li, H. Dani, X. Hu, J. Tang, Y. Chang, and H. Liu. 2017b. Attributed network embedding for learning in a dynamic environment. In *Proc. of the ACM on Conference on Information and Knowledge Management*, pages 387–396. DOI: 10.1145/3132847.3132919 38

Q. Li, Z. Han, and X.-M. Wu. 2018. Deeper insights into graph convolutional networks for semi-supervised learning. In *Proc. of AAAI*, pages 3538–3545. 41, 56

Y. Li, J. Nie, Y. Zhang, B. Wang, B. Yan, and F. Weng. 2010. Contextual recommendation based on text mining. In *Proc. of the 23rd International Conference on Computational Linguistics: Posters*, pages 692–700, Association for Computational Linguistics. 164, 169

J. Liang, S. Gurukar, and S. Parthasarathy. 2018. Mile: A multi-level framework for scalable graph embedding. *ArXiv Preprint ArXiv:1802.09612*. 103, 107, 116

L. Liao, X. He, H. Zhang, and T.-S. Chua. 2018. Attributed social network embedding. *IEEE Transactions on Knowledge and Data Engineering*, 30(12):2257–2270. DOI: 10.1109/tkde.2018.2819980 38

D. Liben-Nowell and J. Kleinberg. 2007. The link-prediction problem for social networks. *Journal of the Association for Information Science and Technology*, 58(7):1019–1031. DOI: 10.1002/asi.20591 3

D. Liben-Nowell and J. Kleinberg. 2008. Tracing information flow on a global scale using internet chain-letter data. *Proc. of the National Academy of Sciences*, 105(12):4633–4638. DOI: 10.1073/pnas.0708471105 172

Y. Lin, S. Shen, Z. Liu, H. Luan, and M. Sun. 2016. Neural relation extraction with selective attention over instances. In *Proc. of ACL*, 1:2124–2133. DOI: 10.18653/v1/p16-1200 145

N. Liu, Q. Tan, Y. Li, H. Yang, J. Zhou, and X. Hu. 2019a. Is a single vector enough? Exploring node polysemy for network embedding. In *Proc. of the 25th ACM SIGKDD International Conference on Knowledge Discovery and Data Mining*, pages 932–940. DOI: 10.1145/3292500.3330967 72, 73

Q. Liu, S. Wu, L. Wang, and T. Tan. 2016. Predicting the next location: A recurrent model with spatial and temporal contexts. In *30th AAAI Conference on Artificial Intelligence*. 169

X. Liu, T. Murata, K.-S. Kim, C. Kotarasu, and C. Zhuang. 2019b. A general view for network embedding as matrix factorization. In *Proc. of the 12th ACM International Conference on Web Search and Data Mining*, pages 375–383. DOI: 10.1145/3289600.3291029 25

S. Lloyd. 1982. Least squares quantization in PCM. *IEEE Transactions on Information Theory*, 28(2):129–137. DOI: 10.1109/tit.1982.1056489 50

L. Lü and T. Zhou. 2011. Link prediction in complex networks: A survey. *Physica A*. DOI: 10.1016/j.physa.2010.11.027 93

Y. Lu, C. Shi, L. Hu, and Z. Liu. 2019. Relation structure-aware heterogeneous information network embedding. In *Proc. of the AAAI Conference on Artificial Intelligence*, 33:4456–4463. DOI: 10.1609/aaai.v33i01.33014456 xxi, 132

H. Ma. 2014. On measuring social friend interest similarities in recommender systems. In *Proc. of the 37th International ACM SIGIR Conference on Research and Development in Information Retrieval*, pages 465–474. DOI: 10.1145/2600428.2609635 148

A. Machanavajjhala, A. Korolova, and A. D. Sarma. 2011. Personalized social recommendations: Accurate or private. *Proc. of the VLDB Endowment*, 4(7):440–450. DOI: 10.14778/1988776.1988780 148

H. Maron, H. Ben-Hamu, H. Serviansky, and Y. Lipman. 2019. Provably powerful graph networks. In *Advances in Neural Information Processing Systems*, pages 2156–2167. 56

A. McCallum, K. Nigam, J. Rennie, and K. Seymore. 2000. Automating the construction of internet portals with machine learning. *Information Retrieval Journal*, 3:127–163. DOI: 10.1023/A:1009953814988 33, 66, 80, 93

M. McPherson, L. Smith-Lovin, and J. M. Cook. 2001. Birds of a feather: Homophily in social networks. *Annual Review of Sociology*. DOI: 10.1146/annurev.soc.27.1.415 90

Q. Mei, D. Cai, D. Zhang, and C. Zhai. 2008. Topic modeling with network regularization. In *Proc. of the 17th International Conference on World Wide Web*, pages 101–110. DOI: 10.1145/1367497.1367512 34, 37

H. Meyerhenke, P. Sanders, and C. Schulz. 2017. Parallel graph partitioning for complex networks. *IEEE Transactions on Parallel and Distributed Systems*, 28(9):2625–2638. DOI: 10.1109/tpds.2017.2671868 115

T. Mikolov, K. Chen, G. Corrado, and J. Dean. 2013a. Efficient estimation of word representations in vector space. In *Proc. of ICIR*. 67, 90, 131, 136, 137, 141, 145

T. Mikolov, I. Sutskever, K. Chen, G. S. Corrado, and J. Dean. 2013b. Distributed representations of words and phrases and their compositionality. In *Proc. of NIPS*, pages 3111–3119. 10, 32, 65, 106, 108, 117, 131, 159, 162

M. Mintz, S. Bills, R. Snow, and D. Jurafsky. 2009. Distant supervision for relation extraction without labeled data. In *Proc. of IJCNLP*, pages 1003–1011. DOI: 10.3115/1690219.1690287 145

S. Mittal. 2016. A survey of techniques for approximate computing. *ACM Computing Surveys (CSUR)*, 48(4):62. DOI: 10.1145/2893356 149

A. Mnih and R. Salakhutdinov. 2007. Probabilistic matrix factorization. In *Advances in NIPS*. 163

F. Monti, M. Bronstein, and X. Bresson. 2017. Geometric matrix completion with recurrent multi-graph neural networks. In *Proc. of NIPS*, pages 3697–3707. 169

S. Myers and J. Leskovec. 2010. On the convexity of latent social network inference. In *Proc. of NIPS*. 172, 173, 186, 187

N. Natarajan and I. S. Dhillon. 2014. Inductive matrix completion for predicting gene-disease associations. *Bioinformatics*, 30(12):i60–i68. DOI: 10.1093/bioinformatics/btu269 31

M. E. Newman. 2001. Clustering and preferential attachment in growing networks. *Physical Review E*, 64(2):025102. DOI: 10.1103/physreve.64.025102 93

M. E. Newman. 2006. Modularity and community structure in networks. *PNAS*. DOI: 10.1073/pnas.0601602103 7, 87, 98

A. Y. Ng, M. I. Jordan, and Y. Weiss. 2002. On spectral clustering: Analysis and an algorithm. In *Proc. of NIPS*, pages 849–856. 7, 48, 50

K. Nowicki and T. A. B. Snijders. 2001. Estimation and prediction for stochastic blockstructures. *Journal of the American Statistical Association*, 96(455):1077–1087. DOI: 10.1198/016214501753208735 98

T. Opsahl and P. Panzarasa. 2009. Clustering in weighted networks. *Social Networks*, 31(2):155–163. DOI: 10.1016/j.socnet.2009.02.002 179

M. Ou, P. Cui, J. Pei, Z. Zhang, and W. Zhu. 2016. Asymmetric transitivity preserving graph embedding. In *Proc. of the 22nd ACM SIGKDD International Conference on Knowledge Discovery and Data Mining*, pages 1105–1114. DOI: 10.1145/2939672.2939751 145

G. Palla, I. Derényi, I. Farkas, and T. Vicsek. 2005. Uncovering the overlapping community structure of complex networks in nature and society. *Nature*, 435(7043):814–818. DOI: 10.1038/nature03607 94, 98

S. Pan, J. Wu, X. Zhu, C. Zhang, and Y. Wang. 2016. Tri-party deep network representation. *Network*, 11(9):12. DOI: 10.1007/s11771-019-4210-8 38

S. Pan, R. Hu, G. Long, J. Jiang, L. Yao, and C. Zhang. 2018. Adversarially regularized graph autoencoder for graph embedding. In *Proc. of IJCAI*, pages 2609–2615. DOI: 10.24963/ijcai.2018/362 41, 49

J. Park, M. Lee, H. J. Chang, K. Lee, and J. Y. Choi. 2019. Symmetric graph convolutional autoencoder for unsupervised graph representation learning. In *Proc. of ICCV*, pages 6519–6528. DOI: 10.1109/iccv.2019.00662 41, 50

F. Pedregosa, G. Varoquaux, A. Gramfort, V. Michel, B. Thirion, O. Grisel, M. Blondel, P. Prettenhofer, R. Weiss, V. Dubourg, et al. 2011. Scikit-learn: Machine learning in python. *JMLR*, 12:2825–2830. 141

W. Pei, T. Ge, and B. Chang. 2014. Max-margin tensor neural network for Chinese word segmentation. In *Proc. of ACL*, pages 293–303. DOI: 10.3115/v1/p14-1028 83

B. Perozzi, R. Al-Rfou, and S. Skiena. 2014. DeepWalk: Online learning of social representations. In *Proc. of SIGKDD*, pages 701–710. DOI: 10.1145/2623330.2623732 5, 6, 10, 14, 15, 21, 29, 34, 37, 50, 60, 67, 80, 87, 88, 101, 104, 108, 126, 141, 152, 163, 183

H. Pham, C. Shahabi, and Y. Liu. 2013. EBM: An entropy-based model to infer social strength from spatiotemporal data. In *Proc. of the ACM SIGMOD International Conference on Management of Data*, pages 265–276. DOI: 10.1145/2463676.2465301 148

H. Pinto, J. M. Almeida, and M. A. Gonçalves. 2013. Using early view patterns to predict the popularity of Youtube videos. In *Proc. of WSDM*. DOI: 10.1145/2433396.2433443 186

A. Pothen, H. D. Simon, and K.-P. Liou. 1990. Partitioning sparse matrices with eigenvectors of graphs. *SIAM Journal on Matrix Analysis and Applications*, 11(3):430–452. DOI: 10.1137/0611030 98

J. Qiu, Y. Dong, H. Ma, J. Li, K. Wang, and J. Tang. 2018. Network embedding as matrix factorization: Unifying DeepWalk, line, pte, and node2vec. In *Proc. of the 11th ACM International Conference on Web Search and Data Mining*, pages 459–467. DOI: 10.1145/3159652.3159706 25

J. Qiu, Y. Dong, H. Ma, J. Li, C. Wang, K. Wang, and J. Tang. 2019. NetSMF: Large-scale network embedding as sparse matrix factorization. In *The World Wide Web Conference*, pages 1509–1520. DOI: 10.1145/3308558.3313446 116

B. Recht, C. Re, S. Wright, and F. Niu. 2011. Hogwild: A lock-free approach to parallelizing stochastic gradient descent. In *Advances in Neural Information Processing Systems*, pages 693–701. 106

S. Rendle, C. Freudenthaler, and L. Schmidt-Thieme. 2010. Factorizing personalized Markov chains for next-basket recommendation. In *Proc. of WWW*. DOI: 10.1145/1772690.1772773 160, 162, 169

S. Riedel, L. Yao, and A. McCallum. 2010. Modeling relations and their mentions without labeled text. In *Proc. of ECML-PKDD*, pages 148–163. DOI: 10.1007/978-3-642-15939-8_10 145

T. Rocktäschel, E. Grefenstette, K. M. Hermann, T. Kočiskỳ, and P. Blunsom. 2015. Reasoning about entailment with neural attention. *ArXiv Preprint ArXiv:1509.06664*. 72

M. G. Rodriguez, D. Balduzzi, and B. Schölkopf. 2011. Uncovering the temporal dynamics of diffusion networks. *ArXiv Preprint ArXiv:1105.0697*. 177

M. G. Rodriguez, J. Leskovec, D. Balduzzi, and B. Schölkopf. 2014. Uncovering the structure and temporal dynamics of information propagation. *Network Science*, 2(1):26–65. DOI: 10.1017/nws.2014.3 172, 174, 177, 180, 186

B. T. C. G. D. Roller. 2004. Max-margin Markov networks. In *Proc. of NIPS*. 83

D. M. Romero, B. Meeder, and J. Kleinberg. 2011. Differences in the mechanics of information diffusion across topics: Idioms, political hashtags, and complex contagion on twitter. In *Proc. of WWW*, pages 695–704, ACM. DOI: 10.1145/1963405.1963503 172

S. T. Roweis and L. K. Saul. 2000. Nonlinear dimensionality reduction by locally linear embedding. *Science*, 290(5500):2323–2326. DOI: 10.1126/science.290.5500.2323 9

T. N. Sainath, B. Kingsbury, V. Sindhwani, E. Arisoy, and B. Ramabhadran. 2013. Low-rank matrix factorization for deep neural network training with high-dimensional output targets. In *Acoustics, Speech and Signal Processing (ICASSP), IEEE International Conference on*, pages 6655–6659. DOI: 10.1109/icassp.2013.6638949 116

K. Saito, R. Nakano, and M. Kimura. 2008. Prediction of information diffusion probabilities for independent cascade model. In *Knowledge-Based Intelligent Information and Engineering Systems*, pages 67–75, Springer. DOI: 10.1007/978-3-540-85567-5_9 174, 186

K. Saito, M. Kimura, K. Ohara, and H. Motoda. 2009. Learning continuous-time information diffusion model for social behavioral data analysis. In *Asian Conference on Machine Learning*, pages 322–337, Springer. DOI: 10.1007/978-3-642-05224-8_25 174, 186

M. J. Salganik, P. S. Dodds, and D. J. Watts. 2006. Experimental study of inequality and unpredictability in an artificial cultural market. *Science*. DOI: 10.1126/science.1121066 171

G. Salton and M. J. McGill. *Introduction to Modern Information Retrieval*. New York: McGraw-Hill, 1986. 23, 94

P. Sanders and C. Schulz. 2011. Engineering multilevel graph partitioning algorithms. In *European Symposium on Algorithms*, pages 469–480, Springer. DOI: 10.1007/978-3-642-23719-5_40 103, 115

A. See, M.-T. Luong, and C. D. Manning. 2016. Compression of neural machine translation models via pruning. *ArXiv Preprint ArXiv:1606.09274*. DOI: 10.18653/v1/k16-1029 117

P. Sen, G. M. Namata, M. Bilgic, L. Getoor, B. Gallagher, and T. Eliassi-Rad. 2008. Collective classification in network data. *AI Magazine*. DOI: 10.1609/aimag.v29i3.2157 3, 20, 33, 49, 80, 93

J. Shang, M. Qu, J. Liu, L. M. Kaplan, J. Han, and J. Peng. 2016. Meta-path guided embedding for similarity search in large-scale heterogeneous information networks. *ArXiv Preprint ArXiv:1610.09769*. 119, 122, 127, 131, 132

D. Shen, X. Zhang, R. Henao, and L. Carin. 2018. Improved semantic-aware network embedding with fine-grained word alignment. In *Proc. of the Conference on Empirical Methods in Natural Language Processing*, pages 1829–1838. DOI: 10.18653/v1/d18-1209 72

C. Shi, X. Kong, Y. Huang, S. Y. Philip, and B. Wu. 2014. Hetesim: A general framework for relevance measure in heterogeneous networks. *IEEE Transactions on Knowledge and Data Engineering*, 26(10):2479–2492. DOI: 10.1109/tkde.2013.2297920 127

C. Shi, Y. Li, J. Zhang, Y. Sun, and S. Y. Philip. 2017. A survey of heterogeneous information network analysis. *IEEE Transactions on Knowledge and Data Engineering*, 29(1):17–37. DOI: 10.1109/TKDE.2016.2598561 119

C. Shi, B. Hu, X. Zhao, and P. Yu. 2018. Heterogeneous information network embedding for recommendation. *IEEE Transactions on Knowledge and Data Engineering*. DOI: 10.1109/tkde.2018.2833443 120, 131, 132

R. Shu and H. Nakayama. 2017. Compressing word embeddings via deep compositional code learning. *ArXiv Preprint ArXiv:1711.01068*. 117

M. Simonovsky and N. Komodakis. 2018. GraphVAE: Towards generation of small graphs using variational autoencoders. In *International Conference on Artificial Neural Networks*, pages 412–422, Springer. DOI: 10.1007/978-3-030-01418-6_41 191

N. Srivastava, G. Hinton, A. Krizhevsky, I. Sutskever, and R. Salakhutdinov. 2014. Dropout: A simple way to prevent neural networks from overfitting. *The Journal of Machine Learning Research*, 15(1):1929–1958. 140

F. M. Suchanek, G. Kasneci, and G. Weikum. 2007. Yago: A core of semantic knowledge. In *Proc. of WWW*, pages 697–706. DOI: 10.1145/1242572.1242667 145

X. Sun, J. Guo, X. Ding, and T. Liu. 2016. A general framework for content-enhanced network representation learning. *ArXiv Preprint ArXiv:1610.02906*. 38, 67

Y. Sun, J. Han, X. Yan, P. S. Yu, and T. Wu. 2011. Pathsim: Meta path-based top-k similarity search in heterogeneous information networks. *Proc. of VLDB*, 4(11):992–1003. DOI: 10.14778/3402707.3402736 119

Y. Sun, B. Norick, J. Han, X. Yan, P. S. Yu, and X. Yu. 2013. Pathselclus: Integrating meta-path selection with user-guided object clustering in heterogeneous information networks. *ACM Transactions on Knowledge Discovery from Data (TKDD)*, 7(3):11. DOI: 10.1145/2513092.2500492 122

M. Surdeanu, J. Tibshirani, R. Nallapati, and C. D. Manning. 2012. Multi-instance multi-label learning for relation extraction. In *Proc. of EMNLP*, pages 455–465. 145

K. S. Tai, R. Socher, and C. D. Manning. 2015. Improved semantic representations from tree-structured long short-term memory networks. In *Proc. of ACL*. DOI: 10.3115/v1/p15-1150 60, 63

J. Tang, J. Zhang, L. Yao, J. Li, L. Zhang, and Z. Su. 2008. Arnetminer: Extraction and mining of academic social networks. In *Proc. of the 14th ACM SIGKDD International Conference on Knowledge Discovery and Data Mining*, pages 990–998. DOI: 10.1145/1401890.1402008 122, 140

J. Tang, M. Qu, and Q. Mei. 2015a. Pte: Predictive text embedding through large-scale heterogeneous text networks. In *Proc. of SIGKDD*. DOI: 10.1145/2783258.2783307 25, 120, 127, 131, 132, 163

J. Tang, M. Qu, M. Wang, M. Zhang, J. Yan, and Q. Mei. 2015b. Line: Large-scale information network embedding. In *Proc. of WWW*. DOI: 10.1145/2736277.2741093 5, 12, 14, 17, 21, 60, 62, 67, 80, 87, 94, 101, 107, 108, 126, 141

L. Tang and H. Liu. 2009. Relational learning via latent social dimensions. In *Proc. of SIGKDD*, pages 817–826, ACM. DOI: 10.1145/1557019.1557109 5, 22, 29, 37, 93, 107

L. Tang and H. Liu. 2011. Leveraging social media networks for classification. *Data Mining and Knowledge Discovery*, 23(3):447–478. DOI: 10.1007/s10618-010-0210-x 10, 14, 15, 20, 21, 22, 29, 37

B. Taskar, D. Klein, M. Collins, D. Koller, and C. D. Manning. 2004. Max-margin parsing. In *Proc. of EMNLP*, 1:3. 83

G. Taubin. 1995. A signal processing approach to fair surface design. In *Proc. of the 22nd Annual Conference on Computer Graphics and Interactive Techniques*, pages 351–358. DOI: 10.1145/218380.218473 41, 45

J. B. Tenenbaum, V. De Silva, and J. C. Langford. 2000. A global geometric framework for nonlinear dimensionality reduction. *Science*, 290(5500):2319–2323. DOI: 10.1126/science.290.5500.2319 9

Y. Tian, H. Chen, B. Perozzi, M. Chen, X. Sun, and S. Skiena. 2019. Social relation inference via label propagation. In *European Conference on Information Retrieval*, pages 739–746, Springer. DOI: 10.1007/978-3-030-15712-8_48 145

O. Tsur and A. Rappoport. 2012. What's in a hashtag?: Content based prediction of the spread of ideas in microblogging communities. In *Proc. of WSDM*, pages 643–652, ACM. DOI: 10.1145/2124295.2124320 186

C. Tu, H. Wang, X. Zeng, Z. Liu, and M. Sun. 2016a. Community-enhanced network representation learning for network analysis. *ArXiv Preprint ArXiv:1611.06645*. xxi, 98

C. Tu, W. Zhang, Z. Liu, and M. Sun. 2016b. Max-margin DeepWalk: Discriminative learning of network representation. In *Proc. of IJCAI*. xxi, 83

C. Tu, H. Liu, Z. Liu, and M. Sun. 2017a. Cane: Context-aware network embedding for relation modeling. In *Proc. of the 55th Annual Meeting of the Association for Computational Linguistics (Volume 1: Long Papers)*, pages 1722–1731. DOI: 10.18653/v1/p17-1158 xxi, 73

C. Tu, Z. Zhang, Z. Liu, and M. Sun. 2017b. TransNet: Translation-based network representation learning for social relation extraction. In *IJCAI*, pages 2864–2870. DOI: 10.24963/ijcai.2017/399 xxi, 145

R. van den Berg, T. N. Kipf, and M. Welling. 2017. Graph convolutional matrix completion. *ArXiv Preprint ArXiv:1706.02263.* 169

A. Van den Oord, S. Dieleman, and B. Schrauwen. 2013. Deep content-based music recommendation. In *Advances in Neural Information Processing Systems*, pages 2643–2651. 177

L. Van Der Maaten. 2014. Accelerating t-SNE using tree-based algorithms. *The Journal of Machine Learning Research*, 15(1):3221–3245. 55

A. Vaswani, N. Shazeer, N. Parmar, J. Uszkoreit, L. Jones, A. N. Gomez, Ł. Kaiser, and I. Polosukhin. 2017. Attention is all you need. In *Advances in Neural Information Processing Systems*, pages 5998–6008. 176

P. Veličković, G. Cucurull, A. Casanova, A. Romero, P. Lio, and Y. Bengio. 2018. Graph attention networks. In *Proc. of ICLR*. 41, 170

E. M. Voorhees et al. 1999. The trec-8 question answering track report. In *Trec*, 99:77–82. DOI: 10.1017/s1351324901002789 7, 109

J. Wallinga and P. Teunis. 2004. Different epidemic curves for severe acute respiratory syndrome reveal similar impacts of control measures. *American Journal of Epidemiology*. DOI: 10.1093/aje/kwh255 171, 172, 173, 187

C. Wang, S. Pan, G. Long, X. Zhu, and J. Jiang. 2017a. MGAE: Marginalized graph autoencoder for graph clustering. In *Proc. of CIKM*, pages 889–898. DOI: 10.1145/3132847.3132967 41, 50

C. Wang, S. Pan, R. Hu, G. Long, J. Jiang, and C. Zhang. 2019a. Attributed graph clustering: A deep attentional embedding approach. In *Proc. of IJCAI*, pages 3670–3676. DOI: 10.24963/ijcai.2019/509 41, 46, 50

D. Wang, P. Cui, and W. Zhu. 2016a. Structural deep network embedding. In *Proc. of KDD*. DOI: 10.1145/2939672.2939753 13, 38, 192

D. Wang, X. Zhang, D. Yu, G. Xu, and S. Deng. 2020. Came: Content-and context-aware music embedding for recommendation. *IEEE Transactions on Neural Networks and Learning Systems*. DOI: 10.1109/tnnls.2020.2984665 72

F. Wang, T. Li, X. Wang, S. Zhu, and C. Ding. 2011. Community discovery using nonnegative matrix factorization. *Data Mining and Knowledge Discovery*. DOI: 10.1007/s10618-010-0181-y 98

H. Wang, J. Wang, J. Wang, M. Zhao, W. Zhang, F. Zhang, X. Xie, and M. Guo. 2017b. GraphGAN: Graph representation learning with generative adversarial nets. *ArXiv Preprint ArXiv:1711.08267.* 191

H. Wang, F. Zhang, M. Hou, X. Xie, M. Guo, and Q. Liu. 2018a. Shine: Signed heterogeneous information network embedding for sentiment link prediction. In *Proc. of WSDM*, pages 592–600, ACM. DOI: 10.1145/3159652.3159666 120, 131, 132

J. Wang, V. W. Zheng, Z. Liu, and K. C.-C. Chang. 2017c. Topological recurrent neural network for diffusion prediction. In *ICDM*, pages 475–484, IEEE. DOI: 10.1109/icdm.2017.57 173, 174, 179, 180, 187

P. Wang, J. Guo, Y. Lan, J. Xu, S. Wan, and X. Cheng. 2015. Learning hierarchical representation model for nextbasket recommendation. In *Proc. of SIGIR*. DOI: 10.1145/2766462.2767694 160, 162

Q. Wang, Z. Mao, B. Wang, and L. Guo. 2017d. Knowledge graph embedding: A survey of approaches and applications. *IEEE Transactions on Knowledge and Data Engineering*, 29(12):2724–2743. DOI: 10.1109/tkde.2017.2754499 145

S. Wang, X. Hu, P. S. Yu, and Z. Li. 2014. MMRate: Inferring multi-aspect diffusion networks with multi-pattern cascades. In *Proc. of SIGKDD*, pages 1246–1255, ACM. DOI: 10.1145/2623330.2623728 186

S. Wang, J. Tang, C. Aggarwal, and H. Liu. 2016b. Linked document embedding for classification. In *Proc. of the 25th ACM International on Conference on Information and Knowledge Management*, pages 115–124. DOI: 10.1145/2983323.2983755 38

S. Wang, C. Aggarwal, J. Tang, and H. Liu. 2017e. Attributed signed network embedding. In *Proc. of the ACM on Conference on Information and Knowledge Management*, pages 137–146. DOI: 10.1145/3132847.3132905 38, 144

S. Wang, J. Tang, C. Aggarwal, Y. Chang, and H. Liu. 2017f. Signed network embedding in social media. In *Proc. of SDM*. DOI: 10.1137/1.9781611974973.37 144

X. Wang, P. Cui, J. Wang, J. Pei, W. Zhu, and S. Yang. 2017g. Community preserving network embedding. In *Proc. of AAAI*. 88, 98

X. Wang, X. He, Y. Cao, M. Liu, and T.-S. Chua. 2019b. KGAT: Knowledge graph attention network for recommendation. In *Proc. of SIGKDD*, pages 950–958. DOI: 10.1145/3292500.3330989 170

X. Wang, X. He, M. Wang, F. Feng, and T.-S. Chua. 2019c. Neural graph collaborative filtering. In *Proc. of SIGIR*, pages 165–174. DOI: 10.1145/3331184.3331267 170

Y. Wang, H. Shen, S. Liu, J. Gao, and X. Cheng. 2017h. Cascade dynamics modeling with attention-based recurrent neural network. In *Proc. of IJCAI*. DOI: 10.24963/ijcai.2017/416 187

Z. Wang, C. Chen, and W. Li. 2018b. Attention network for information diffusion prediction. In *Proc. of WWW*. DOI: 10.1145/3184558.3186931 187

Z. Wang, C. Chen, and W. Li. 2018c. A sequential neural information diffusion model with structure attention. In *Proc. of CIKM*. DOI: 10.1145/3269206.3269275 187

S. Wasserman and K. Faust. 1994. *Social Network Analysis: Methods and Applications*, vol. 8. Cambridge University Press. DOI: 10.1017/cbo9780511815478 123

D. J. Watts and P. S. Dodds. 2007. Influentials, networks, and public opinion formation. *Journal of Consumer Research*, 34(4):441–458. DOI: 10.1086/518527 171

L. Weng, F. Menczer, and Y.-Y. Ahn. 2014. Predicting successful memes using network and community structure. In *8th International AAAI Conference on Weblogs and Social Media*, pages 535–544. 186

F. Wu, T. Zhang, A. H. d. Souza Jr, C. Fifty, T. Yu, and K. Q. Weinberger. 2019a. Simplifying graph convolutional networks. In *Proc. of ICML*, pages 6861–6871. 41

L. Wu, C. Quan, C. Li, Q. Wang, B. Zheng, and X. Luo. 2019b. A context-aware user-item representation learning for item recommendation. *ACM Transactions on Information Systems (TOIS)*, 37(2):1–29. DOI: 10.1145/3298988 72

M. Xie, H. Yin, H. Wang, F. Xu, W. Chen, and S. Wang. 2016. Learning graph-based poi embedding for location-based recommendation. In *Proc. of the 25th ACM International on Conference on Information and Knowledge Management*, pages 15–24. DOI: 10.1145/2983323.2983711 169

K. Xu, W. Hu, J. Leskovec, and S. Jegelka. 2018. How powerful are graph neural networks? *ArXiv Preprint ArXiv:1810.00826*. 56

L. Xu, X. Wei, J. Cao, and P. S. Yu. 2017. Embedding of embedding (EOE): Joint embedding for coupled heterogeneous networks. In *Proc. of WSDM*, pages 741–749, ACM. DOI: 10.1145/3018661.3018723 120, 131, 132

C. Yang and Z. Liu. 2015. Comprehend DeepWalk as matrix factorization. *ArXiv Preprint ArXiv:1501.00358*. 15

C. Yang, Z. Liu, D. Zhao, M. Sun, and E. Y. Chang. 2015. Network representation learning with rich text information. In *Proc. of IJCAI*. xxi, 14, 17, 21, 38, 49, 50, 67, 77, 163

C. Yang, M. Sun, Z. Liu, and C. Tu. 2017a. Fast network embedding enhancement via high order proximity approximation. In *Proc. of IJCAI*. DOI: 10.24963/ijcai.2017/544 xxi, 25

C. Yang, M. Sun, W. X. Zhao, Z. Liu, and E. Y. Chang. 2017b. A neural network approach to jointly modeling social networks and mobile trajectories. *ACM Transactions on Information Systems (TOIS)*, 35(4):1–28. DOI: 10.1145/3041658 xxi, 170

C. Yang, M. Sun, H. Liu, S. Han, Z. Liu, and H. Luan. 2018a. Neural diffusion model for microscopic cascade prediction. *ArXiv Preprint ArXiv:1812.08933*. DOI: 10.1109/tkde.2019.2939796 xxi, 187

J. Yang and J. Leskovec. 2012. Community-affiliation graph model for overlapping network community detection. In *Proc. of ICDM*. DOI: 10.1109/icdm.2012.139 94, 98, 103, 121, 124

J. Yang and J. Leskovec. 2013. Overlapping community detection at scale: A nonnegative matrix factorization approach. In *Proc. of WSDM*. DOI: 10.1145/2433396.2433471 3, 98

J. Yang, J. McAuley, and J. Leskovec. 2013. Community detection in networks with node attributes. In *Proc. of ICDM*. DOI: 10.1109/icdm.2013.167 87

L. Yang, Y. Guo, and X. Cao. 2018b. Multi-facet network embedding: Beyond the general solution of detection and representation. In *32nd AAAI Conference on Artificial Intelligence*. 72, 73

Z. Yang, W. W. Cohen, and R. Salakhutdinov. 2016. Revisiting semi-supervised learning with graph embeddings. *ArXiv Preprint ArXiv:1603.08861*. 83

J. Ye, Z. Zhu, and H. Cheng. 2013. What's your next move: User activity prediction in location-based social networks. In *Proc. of the SIAM International Conference on Data Mining*. DOI: 10.1137/1.9781611972832.19 169

M. Ye, P. Yin, W.-C. Lee, and D.-L. Lee. 2011. Exploiting geographical influence for collaborative point-of-interest recommendation. In *Proc. of the 34th International ACM SI-GIR Conference on Research and Development in Information Retrieval*, pages 325–334. DOI: 10.1145/2009916.2009962 168

H. Yin, Y. Sun, B. Cui, Z. Hu, and L. Chen. 2013. Lcars: A location-content-aware recommender system. In *Proc. of the 19th ACM SIGKDD International Conference on Knowledge Discovery and Data Mining*, pages 221–229. DOI: 10.1145/2487575.2487608 168

R. Ying, R. He, K. Chen, P. Eksombatchai, W. L. Hamilton, and J. Leskovec. 2018. Graph convolutional neural networks for web-scale recommender systems. In *Proc. of SIGKDD*, pages 974–983, ACM. DOI: 10.1145/3219819.3219890 170

Z. Ying, D. Bourgeois, J. You, M. Zitnik, and J. Leskovec. 2019. GNNExplainer: Generating explanations for graph neural networks. In *Advances in Neural Information Processing Systems*, pages 9244–9255. 192

J. You, B. Liu, Z. Ying, V. Pande, and J. Leskovec. 2018. Graph convolutional policy network for goal-directed molecular graph generation. In *Advances in Neural Information Processing Systems*, pages 6410–6421. 192

H.-F. Yu, P. Jain, and I. S. Dhillon. 2014. Large-scale multi-label learning with missing labels. In *Proc. of ICML*. 31, 32

L. Yu, P. Cui, F. Wang, C. Song, and S. Yang. 2015. From micro to macro: Uncovering and predicting information cascading process with behavioral dynamics. In *ICDM*. DOI: 10.1109/icdm.2015.79 171, 172, 186

J. Yuan, N. Gao, L. Wang, and Z. Liu. 2018. MultNet: An efficient network representation learning for large-scale social relation extraction. In *International Conference on Neural Information Processing*, pages 515–524, Springer. DOI: 10.1007/978-3-030-04182-3_45 145

Q. Yuan, G. Cong, Z. Ma, A. Sun, and N. M. Thalmann. 2013. Time-aware point-of-interest recommendation. In *Proc. of the 36th International ACM SIGIR Conference on Research and Development in Information Retrieval*, pages 363–372. DOI: 10.1145/2484028.2484030 169

Q. Yuan, G. Cong, and C.-Y. Lin. 2014a. Com: A generative model for group recommendation. In *Proc. of the 20th ACM SIGKDD International Conference on Knowledge Discovery and Data Mining*, pages 163–172. DOI: 10.1145/2623330.2623616 148

Q. Yuan, G. Cong, and A. Sun. 2014b. Graph-based point-of-interest recommendation with geographical and temporal influences. In *Proc. of the 23rd ACM International Conference on Conference on Information and Knowledge Management*, pages 659–668. DOI: 10.1145/2661829.2661983 169

S. Yuan, X. Wu, and Y. Xiang. 2017. SNE: Signed network embedding. In *Pacific–Asia Conference on Knowledge Discovery and Data Mining*, pages 183–195, Springer. DOI: 10.1007/978-3-319-57529-2_15 144

W. W. Zachary. 1977. An information flow model for conflict and fission in small groups. *JAR*. DOI: 10.1086/jar.33.4.3629752 97

N. S. Zekarias Kefato and A. Montresor. 2017. Deepinfer: Diffusion network inference through representation learning. In *Proc. of the 13th International Workshop on Mining and Learning with Graphs (MLG)*, p. 5. 173, 174

D. Zhang, J. Yin, X. Zhu, and C. Zhang. 2016. Homophily, structure, and content augmented network representation learning. In *IEEE 16th International Conference on Data Mining (ICDM)*, pages 609–618. DOI: 10.1109/icdm.2016.0072 37

H. Zhang, I. King, and M. R. Lyu. 2015. Incorporating implicit link preference into overlapping community detection. In *Proc. of AAAI*. 95

J. Zhang, C. Xia, C. Zhang, L. Cui, Y. Fu, and S. Y. Philip. 2017a. Bl-mne: Emerging heterogeneous social network embedding through broad learning with aligned autoencoder. In *Proc. of ICDM*, pages 605–614, IEEE. DOI: 10.1109/icdm.2017.70 120, 131

J. Zhang, Y. Dong, Y. Wang, J. Tang, and M. Ding. 2019a. Prone: Fast and scalable network representation learning. In *IJCAI*, pages 4278–4284. DOI: 10.24963/ijcai.2019/594 116

J. Zhang, X. Shi, S. Zhao, and I. King. 2019b. STAR-GCN: Stacked and reconstructed graph convolutional networks for recommender systems. In S. Kraus, Ed., *Proc. of IJCAI*, pages 4264–4270. DOI: 10.24963/ijcai.2019/592 170

J.-D. Zhang, C.-Y. Chow, and Y. Li. 2014. Lore: Exploiting sequential influence for location recommendations. In *Proc. of the 22nd ACM SIGSPATIAL International Conference on Advances in Geographic Information Systems*, pages 103–112. DOI: 10.1145/2666310.2666400 169

X. Zhang, W. Chen, F. Wang, S. Xu, and B. Xu. 2017b. Towards compact and fast neural machine translation using a combined method. In *Proc. of the Conference on Empirical Methods in Natural Language Processing*, pages 1475–1481. DOI: 10.18653/v1/d17-1154 117

X. Zhang, H. Liu, Q. Li, and X.-M. Wu. 2019c. Attributed graph clustering via adaptive graph convolution. In *Proc. of IJCAI*, pages 4327–4333. DOI: 10.24963/ijcai.2019/601 50

Z. Zhang, P. Cui, H. Li, X. Wang, and W. Zhu. 2018a. Billion-scale network embedding with iterative random projection. In *IEEE International Conference on Data Mining (ICDM)*, pages 787–796. DOI: 10.1109/icdm.2018.00094 116

Z. Zhang, C. Yang, Z. Liu, M. Sun, Z. Fang, B. Zhang, and L. Lin. 2018b. Cosine: Compressive network embedding on large-scale information networks. *ArXiv Preprint ArXiv:1812.08972*. DOI: 10.1109/tkde.2020.3030539 xxi, 117

K. Zhao, G. Cong, Q. Yuan, and K. Q. Zhu. 2015a. Sar: A sentiment-aspect-region model for user preference analysis in geo-tagged reviews. In *IEEE 31st International Conference on Data Engineering*, pages 675–686. DOI: 10.1109/icde.2015.7113324 164, 169

Q. Zhao, M. A. Erdogdu, H. Y. He, A. Rajaraman, and J. Leskovec. 2015b. Seismic: A self-exciting point process model for predicting tweet popularity. In *Proc. of SIGKDD*. DOI: 10.1145/2783258.2783401 171, 186

W. X. Zhao, N. Zhou, W. Zhang, J.-R. Wen, S. Wang, and E. Y. Chang. 2016. A probabilistic lifestyle-based trajectory model for social strength inference from human trajectory data. *ACM Transactions on Information Systems (TOIS)*, 35(1):8. DOI: 10.1145/2948064 168

A. Zheng, C. Feng, F. Yang, and H. Zhang. 2019. EsiNet: Enhanced network representation via further learning the semantic information of edges. In *IEEE 31st International Conference on Tools with Artificial Intelligence (ICTAI)*, pages 1011–1018. DOI: 10.1109/ictai.2019.00142 145

V. W. Zheng, Y. Zheng, X. Xie, and Q. Yang. 2010. Collaborative location and activity recommendations with GPS history data. In *Proc. of the 19th International Conference on World Wide Web*, pages 1029–1038, ACM. DOI: 10.1145/1772690.1772795 169

Y. Zheng. 2015. Trajectory data mining: An overview. *ACM Transactions on Intelligent Systems and Technology (TIST)*, 6(3):29. DOI: 10.1145/2743025 148

Y. Zheng, L. Zhang, Z. Ma, X. Xie, and W.-Y. Ma. 2011. Recommending friends and locations based on individual location history. *ACM Transactions on the Web (TWEB)*, 5(1):5. DOI: 10.1145/1921591.1921596 148

C. Zhou, Y. Liu, X. Liu, Z. Liu, and J. Gao. 2017. Scalable graph embedding for asymmetric proximity. In *Proc. of the 31st AAAI Conference on Artificial Intelligence*, pages 2942–2948. 145

J. Zhou, G. Cui, Z. Zhang, C. Yang, Z. Liu, L. Wang, C. Li, and M. Sun. 2018. Graph neural networks: A review of methods and applications. *ArXiv Preprint ArXiv:1812.08434*. 39, 57

N. Zhou, W. X. Zhao, X. Zhang, J.-R. Wen, and S. Wang. 2016. A general multi-context embedding model for mining human trajectory data. *IEEE Transactions on Knowledge and Data Engineering*. DOI: 10.1109/tkde.2016.2550436 168

T. Zhou, L. Lü, and Y.-C. Zhang. 2009. Predicting missing links via local information. *The European Physical Journal B*, 71(4):623–630. DOI: 10.1140/epjb/e2009-00335-8 94

J. Zhu, A. Ahmed, and E. P. Xing. 2012. Medlda: Maximum margin supervised topic models. *JMLR*, 13(1):2237–2278. DOI: 10.1145/1553374.1553535 78, 83

Y. Zhu, Y. Xu, F. Yu, Q. Liu, S. Wu, and L. Wang. 2020. Deep graph contrastive representation learning. *ArXiv Preprint ArXiv:2006.04131*. 191

Authors' Biographies

CHENG YANG

Cheng Yang is an assistant professor in the School of Computer Science, Beijing University of Posts and Telecommunications, China. He received his B.E. and Ph.D. degrees in Computer Science from Tsinghua University in 2014 and 2019, respectively. His research interests include network representation learning, social computing, and natural language processing. He has published more than 20 papers in top-tier conferences and journals including AAAI, ACL, ACM TOIS, and IEEE TKDE.

ZHIYUAN LIU

Zhiyuan Liu is an associate professor in the Department of Computer Science and Technology, Tsinghua University, China. He got his B.E. in 2006 and his Ph.D. in 2011 from the Department of Computer Science and Technology, Tsinghua University. His research interests are natural language processing and social computation. He has published over 60 papers in international journals and conferences, including IJCAI, AAAI, ACL, and EMNLP, and received more than 10,000 citations according to Google Scholar.

CUNCHAO TU

Cunchao Tu is a postdoc in the Department of Computer Science and Technology, Tsinghua University. He got his B.E. and Ph.D. in 2013 and 2018 from the Department of Computer Science and Technology, Tsinghua University. His research interests include network representation learning, social computing, and legal intelligence. He has published over 20 papers in international journals and conferences including IEEE TKDE, AAAI, ACL, and EMNLP.

CHUAN SHI

Chuan Shi is a professor in the School of Computer Sciences of Beijing University of Posts and Telecommunications. His main research interests include data mining, machine learning, and big data analysis. He has published more than 100 refereed papers, including top journals and conferences in data mining, such as IEEE TKDE, ACM TIST, KDD, WWW, AAAI, and IJCAI.

MAOSONG SUN

Maosong Sun is a professor of the Department of Computer Science and Technology, and the Executive Vice President of Institute of Artificial Intelligence at Tsinghua University. His research interests include natural language processing, internet intelligence, machine learning, social computing, and computational education. He was elected as a foreign member of the Academia Europaea in 2020. He has published more than 200 papers in top-tier conferences and journals and received more than 15,000 citations according to Google Scholar.

Printed in the United States
by Baker & Taylor Publisher Services